guesstimation 2.0

Lawrence Weinstein

¿ ¿

gue**??**timation 2.0

Solving Today's Problems
on the Back of a Napkin

Princeton University Press   Princeton and Oxford

Published by Princeton University Press, 41 William Street, Princeton, New Jersey 08540

In the United Kingdom: Princeton University Press, 6 Oxford Street, Woodstock, Oxfordshire OX20 1TW

Library of Congress Cataloging-in-Publication Data

Weinstein, Lawrence, 1960–

Guesstimation 2.0: solving today's problems on the back of a napkin / Lawrence Weinstein.

   p. cm.

Includes bibliographical references and index.

ISBN 978-0-691-15080-2 (pbk.: alk. paper) 1. Estimation theory—Miscellanea.   2. Problem solving—Miscellanea.   I. Title.

QA276.8.W45 2012

519.5′44—dc23

2012011182

British Library Cataloging-in-Publication Data is available

This book has been composed in Minion Pro with Scala Sans display

Printed on acid-free paper. ∞

press.princeton.edu

Typeset by S R Nova Pvt Ltd, Bangalore, India

Printed in the United States of America

10 9 8 7 6 5 4 3 2 1

To my parents, who taught me to love learning

Contents

Appendix A
Dealing with Large Numbers 341

Appendix B
Pegs to Hang Things On 347

Acknowledgments

I would like to thank the many people who helped me write this book (and apologize to those I inadvertantly omit).

First and foremost, I thank the Physics Department of Old Dominion University, for providing an intellectually stimulating environment, for encouraging my interest in estimation, and for tolerating the time I spend writing. I thank my friend and colleague, John Adam, for many enjoyable discussions about the book and for helping to start my writing career by co-authoring *Guesstimation*. I thank my editor at Princeton University Press, Vickie Kearn, for being helpful, friendly, and supportive during all phases of the project (and for being so patient when the manuscript was a bit late). Thanks also to the folks at the *Physics Teacher*, who provide space for my monthly estimation column (from which some of the problems in this book are modified). I am grateful to the many colleagues and students who suggested interesting questions, both for the column and for this book.

I thank my wife and children, Carol, Lee, and Rachel, for their specific suggestions and corrections, for their general support of this project, for their toleration when I get that vacant look in my eyes and start mumbling exponents, and for making life fun and worthwhile.

Lastly, this book is dedicated to my parents. I thank my father for encouraging my love of mathematics, and I thank my mother for showing that writing books is not that scary. When I asked my mother if she had to be extra careful to be absolutely correct because she was writing a *book*, she laughed and responded that she had far more freedom to write what she wanted in a book than in a magazine article.

Any errors or omissions in this book are, of course, all mine.

Preface

How much is this book worth to you?

That is a typical estimation question. You are not looking for a precise answer. Your answer will help determine a future action (e.g., buying the book or not). And you can answer it in several different ways. For example, you could consider the value of finally knowing the answer to that age-old question, "paper or plastic?" once and for all, so that you no longer need to agonize over it at the grocery store. You could consider the value of improving your estimation abilities, either as an enjoyable pastime or as a useful business skill. You could consider the value of whiling away a few hours, thinking about interesting problems and reading amusing solutions. Lastly, you could consider the value of acing that job interview with Google. If this book does help you get that high-flying job, please let me know (I promise not to charge you more for it).

If you estimate that the value of this book is more than its cover price, then I hope you will buy it. If you estimate that its value is less than its price, then I hope you will not buy it (or that you will be pleasantly surprised when reading it).

This book continues where the popular and widely reviewed *Guesstimation* leaves off. We'll look at practical questions such as the value of the solar panel on a Prius and how far we should walk to recycle that water bottle. We'll look at impractical questions such as how far a mousetrap-powered car could travel and whether there are more brains or air in a movie theater. We'll develop the ability to see the news and identify the numerical bovine coprolites (BS, that is).

In addition to learning how to estimate for fun and profit, we'll become a lot more familiar with big numbers. These numbers are all around us, and some of them are even important. There are a billion people in China. The federal budget deficit is a trillion dollars. The president was paid a million dollars. After a while all these zillions sound the same. However, a trillion is a thousand times bigger than a billion, which is a thousand times bigger than a million. A factor of a thousand is a lot. If you make $50,000 a year, then you probably don't worry too much about spending $50 for something important. Similarly, if the federal budget deficit is $1 trillion, then we probably shouldn't worry too much about spending $1 billion for something important (assuming that we can agree on what is important).

Randall Munroe of xkcd.com, describes in a web comic the difference between a million and a billion more graphically.

DEAR NEWS ORGANIZATIONS: STOP GIVING LARGE NUMBERS WITHOUT CONTEXT OR PROPER COMPARISON.

THE DIFFERENCE BETWEEN A MILLION AND A BILLION IS THE DIFFERENCE BETWEEN ME HAVING A SIP OF WINE AND 30 SECONDS WITH YOUR DAUGHTER, AND A BOTTLE OF GIN AND A NIGHT WITH HER.

Recognizing that a billion is *much* larger than a million can help you avoid some embarrassing errors. For example, in May 2010 the Dow Jones Industrial average plunged one thousand points in just a few minutes. According to the *New York Times*, "Federal officials fielded rumors that the culprit was a single stock, a single institution or execution system, a $16 billion trade that should have been $16 million" [1]. Oops.

By the way, Google really does use questions like the following in job interviews. How many golf balls can fit in a school bus? How many piano tuners are there in the whole world? [2, 3]. If you can answer questions like these, you will be more attractive to companies like Google because it shows that you are a flexible thinker, are willing to attack imprecise questions, and can apply your knowledge to real-world questions.

And you will finally know the answer, if not to the question of "life, the universe, and everything" [4], then at least to the question of "paper or plastic?"

gue??timation 2.0

How to Solve Problems

Chapter 1

¿ ¿

Why estimate?

Because we need to know an answer so we can decide what to do. If we're deciding whether to drive to Virginia Beach for the weekend, we need to estimate how long it will take. If we're deciding whether to get a solar panel on our new car, we need to estimate how much gas it will save. If we're deciding between paper and plastic, we need to estimate how much plastic is in the bags we use each year. All these questions (and many more) lead to actions. If the drive is too long, we won't go to Virginia Beach. If the solar panel generates enough energy, we'll buy that option. If we only use a few pounds of plastic bags in a year, we won't agonize over the choice.

We are not looking to determine the answers to these questions precisely. We just need to determine them well enough to decide on an action. This means answers fall into one of the three "Goldilocks" categories:

1. too big
2. too small
3. just right

If the answer is too big or too small, then the action is obvious. If the driving time is about 1 hour (too small) then we will drive to Virginia Beach for the weekend. If the driving time is about 10 hours (too big) then we won't. Only if the driving time is in between (just right), do we need more information (such as how much traffic to expect). Similarly, if the solar panel option on the new car will save about 500 gallons of gasoline each year (too big), then we will buy the solar panel. If the solar panel option will only save about 5 gallons of gasoline each year (too small), then we will not buy it. If it will save an intermediate amount, then we need to do more work to get a more precise answer. Similarly, if we use 1 pound (too small) or 1,000 pounds (too big) of plastic bags a year, then our "paper or plastic" decision is easy. Only if the answer is intermediate (just right) do we need to do more work to determine the exact relative environmental impact of plastic and paper bags. However, the extra work to achieve that level of precision is way beyond the scope of this book.

In this book, we will only try to estimate answers to within a factor of ten, as that will be good enough for most questions.

How do we estimate? There are three steps and one technique to make it easier. First, write down the answer. Second, if necessary, break the problem down into small pieces. Third, recombine the pieces. The final step is to compare our answer to the Goldilocks categories. The technique is to establish boundaries for the answer rather than to estimate it directly.

Let's look at this in more detail.

Step 1: Write down the answer to within a factor of ten.[*] Many problems are simple enough so that we can answer them immediately (within a factor of ten). How long does it take a pumpkin to fall from a ten-story building? How much does a new compact car cost? How long does it take to fly across country? All of these can be answered immediately (10 seconds, $20,000, and 6 hours). However, you need to remember one principle.

Dare to be imprecise!

None of those answers is precisely correct. Precise answers take a lot more time. However, most questions

---

[*] A "factor of ten" refers to the number of zeros in the answer. Is the answer 1, 10, 100, or 1,000?

do not need persnickety precision. If you want to buy a car and can only afford to spend $5,000, then you cannot buy a new car. If you can afford to spend $100,000, then you can buy a new car. Only if you have an intermediate amount, for example $15,000, do you need to know the cost of a new car more precisely.

Of course, if all problems were this easy, you would not bother with this book. If the problem is more complicated than that, we need to simplity it.

Break down the problem

Step 2: Break the problem into smaller pieces. Write down the answer to each of these pieces to within a factor of ten. If the smaller pieces are still too complicated, break them down further. Then we

Recombine the pieces

Step 3: Take the estimates of each piece and multiply them together to get the overall estimate. For example, if we want to estimate the total time that all college students talk on their cell phones, we would multiply our estimate of the number of college students by our estimate of the time each student spends on the phone.

Estimate by bounding

Technique: It is often easier and more reliable to estimate upper and lower limits for something than to estimate the quantity directly. Consider estimating the amount of time each day that the average college student spends talking or texting on his or her cell phone. We don't know if it is 10, 27, 34 or 62 minutes. However, we can confidently claim that it is more than 2 minutes and less than 3 hours (or 200 minutes).[*]

Now we just need to convert those bounds to an estimate. The way to do it is to take an average. But we won't just add the upper and lower bounds and divide by two. That would give us an estimate (100 minutes) that is a factor of two lower than our upper limit, but a factor of fifty greater than our lower limit. Because our goal is to estimate an answer within a factor of ten, we could be way off. (Especially after multiplying the estimates of each piece together.)

The solution is to take the geometric mean. The geometric mean of two numbers is the square root of their product. For example, the geometric mean of 5 and 20 is 10, because $10 = \sqrt{(5 \times 20)}$. However, we won't calculate square roots on the back of a cocktail napkin. That would be too ostentatious and needlessly precise.

To take the *approximate* geometric mean of any two numbers, just average their coefficients and average their exponents.[†] In the phone case, the geometric mean of 2 (2×10^0) minutes [‡] and 200 (2×10^2) minutes is 20 (2×10^1) minutes, because 1 is the average of the exponents 0 and 2. Similarly, the geometric mean of 2×10^{15} and 6×10^3 is about 4×10^9 (because $4 = \frac{2+6}{2}$ and $9 = \frac{15+3}{2}$).

If the sum of the exponents is an odd number, the procedure becomes a little more complicated. Then you should round the resulting exponent down, and

[*] Yes, I know that $3 \times 60 = 180$, not 200. But 200 is close enough for this book.

[†] We use coefficients and exponents to describe numbers in scientific notation. The exponent is the power of ten and the coefficient is the number (between 1 and 9.99) that multiplies the power of ten. If you are not familiar with this notation, please check appendix A ("Dealing with Large Numbers") and then come right back. We'll wait for you here.

[‡] Any number raised to the 0th power is 1.

multiply the final answer by three. Therefore, the geometric mean of 1 and 10^3 is $3 \times 10^1 = 30$.

Compare the answer

After we estimate the answer to a question, there is always one final step. Presumably, we attacked that particular question because the answer might mean something or might guide our actions. However, answers only mean something when compared to something else. Therefore, we need to compare the answer to a reasonable (or unreasonable) standard. Only then can we determine which of the three Goldilocks categories the answer falls into.

Failures of estimation

Lastly, we need to think about when estimation fails. There are three types of failure. First, we may badly over- or underestimate a quantity. However, if we bound our estimates carefully, we should be able to recognize possible problems. For example, if our lower and upper bounds are 10^2 and 10^3, then we can be pretty confident that our estimate of 3×10^2 is reasonably good. If our lower and upper bounds are instead 1 and 10^5, then our estimate is the same, but it has a much larger margin of error.

Second, we may choose the wrong model for our estimate. If we try to estimate the number of automobile accidents per year in the United States, we would get different answers if we assumed that they were mostly caused by teenagers, by senior citizens, by distracted parents, or by drunk drivers.

Finally, we may choose a nonlinear problem. Typically, we assume that if we do something and get a certain response, then two of us doing it will get twice the response (and 3×10^8 Americans will get 3×10^8 more response). This can fail spectacularly.

For example, if I stand on a diving board, it will bend a few inches. If you join me on the board, then it will bend about twice as much. But if twenty of us get on the diving board, it will not bend; it will break.

Similarly, if one car on the highway can transport one person 60 miles in an hour, then one hundred cars on the same highway can transport one hundred people 60 miles in that same hour. However, 10^4 cars on the same highway cannot transport 10^4 people 60 miles in an hour because they will cause a huge traffic jam.

These situations are rare, but we need to watch out for them.

Let's get started.

Example 1: Paving with pizza boxes

How much area could we cover with all the pizza boxes used by Americans in one year? This is too complicated to estimate directly, so we need to break the problem into smaller pieces. Let's estimate the number of pizza boxes we each use a year and the area of each pizza box.

How many pizzas does the average American order each year? We may not know the average, but we can set upper and lower bounds. It is more than one and less than one hundred. Thus, we will estimate that the average is the geometric mean, which is ten pizzas per year. Another approach would be to estimate the number of times per week or per month that we get pizza in a pizza box (either take-out or delivered). If we have pizza once a week and share a pie with a friend, that makes our share twenty-five pizzas per year.

Estimating the area is easier. A large pizza is 18 inches in diameter (or almost 0.5 m). The pizza box will be a slightly larger rectangle with an area of one-quarter square meter ($0.25\,\mathrm{m}^2$). Now all we need to do to is multiply.[*]

$$A = (3 \times 10^8 \text{ Americans}) \times \left(10\frac{\text{pizzas}}{\text{Am-year}} \right)$$
$$\times \left(0.25\frac{\text{m}^2}{\text{pizza}} \right)$$
$$= 10^9\,\mathrm{m}^2 = 10^3\,\mathrm{km}^2 = 500\,\mathrm{mi}^2$$

And now for the final step, comparing the answer. This area is about the area of a large city such as New York City, Los Angeles, or Virginia Beach.[†] That is definitely a lot of pizza boxes.

Example 2: Playing video games

How much time did Americans spend playing video games last year? How many lifetimes?

Again, this question is too complicated to just write down the answer. We will need to break it down into smaller pieces. Let's estimate the percentage of people who play video games and the average time they each spend doing so. Probably about half of all Americans between the ages of ten and forty play video games. It is certainly more than 10% and less than 100%, so we can take the geometric mean and use 30% (or we can just use 50%). We might not know how much time the average gamer spends at his or her console every day. However, we can confidently assume that it is more than 10 minutes and less than 10 hours.[‡] Taking the

---

[*] We do need to know that the U.S. population is approximately 300 million (3×10^8). That is another one of the few numbers we need to memorize in order to make good estimates. Feel free to write it on your shirt cuff or wrist.

[†] That progression is an example of bathos, just like "For God, for country, and for Yale."[5]

[‡] We might be wrong, but we would be *confidently* wrong.

geometric mean of 10 and 600 minutes gives about 1 hour per day.[*]

Is this a reasonable average? It is probably too low for high school and college students and probably too high for thirty-somethings. On the other hand, the answer is likely to be between 0.5 and 2 hours and is almost certain to be between 0.25 and 4 hours, so our estimate will not be too far off.

About one-third to one-half of the population will be between ten and forty years old (as that is thirty years out of our average lifespan of eighty years). This means that, for the general population, the average amount of time per day spent playing video games is

$$t = 1\frac{\text{hour}}{\text{day}} \times 0.5 \times \frac{30\,\text{years}}{80\,\text{years}} = 0.2\frac{\text{hour}}{\text{day}}.$$

Now we need to calculate the total time spent playing video games by all Americans. To do this, we need to multiply the number of hours per day by the number of days in a year and by the total number of Americans.

$$T = 0.2\frac{\text{hour}}{\text{day}} \times 400\frac{\text{days}}{\text{year}} \times (3 \times 10^8\,\text{Americans})$$

$$= 2 \times 10^{10}\,\text{hours/year}$$

That is a lot of time! In fact, that is too much. I do not have a clear idea of how large 10 billion hours is. In order to perform the final step and compare this answer to other times, we first need to convert it to more appropriate units.

---

[*] Yes, yes. The geometric mean of 10 and 600 is 77.45. However, given that neither 10 nor 600 is precisely determined, we won't precisely determine their geometric mean. If our answer turns out to be "just right," then we can worry about fine-tuning it.

Let's see. At 24 hours per day and 400 days per year, there are 10^4 hours per year.[*] That means that 10^{10} hours equals 10^6 years. Wow, that is still a lot of time.

Now let's convert this to lifetimes. At eighty years per lifetime, 10^6 years equals 10^4 lifetimes. However we look at it, it is a tremendous amount of time.

Well, no. The amount of time became huge because we multiplied it by the population of the United States. Let's make a different comparison. The daily time spent playing video games, when averaged over all Americans, is only 0.2 hours. That is 0.2 hours out of 24, or only 1%. Even the people actually playing video games are spending only one hour per day, which is just 4% of their time.

Thus, it is important both to estimate the number correctly and to make appropriate comparisons.

[*] Yes, I know that there are 365 days per year. Would you rather do the arithmetic with 400 or with 365?

General Questions

Chapter 2

Here are some of life's important questions, ranging from the total length of all our toilet paper through the literary efforts of a million monkeys to the amount of pee in our swimming pool. We'll start slowly by estimating length, then area, then volume, and then more intricate questions.

Who unrolled the toilet paper?

If all of the toilet paper used in the United States in an entire year were rolled out, how far would it stretch?

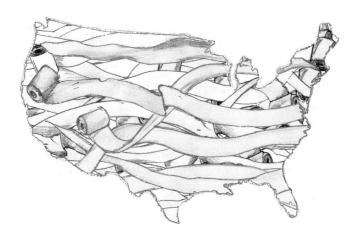

HINT: How much toilet paper do you use every day?

HINT: There are 3×10^8 people in the United States.

ANSWER: To estimate this, we need to break it down into how much toilet paper each American uses daily, how many Americans there are, and how many days there are in a year.

There are two possible approaches to estimating individual toilet paper usage. We can estimate how much toilet paper we use each day, or we can estimate how many rolls of toilet paper we use a month (or a year). I sit on the toilet about once a day and use about ten sheets of toilet paper each time.[*] However, we should probably account for the different waste disposal plumbing and toilet paper requirements of the different genders. Let's double this to twenty sheets per day.[†] Now let's compare this with the alternative estimate. My wife and I replace the toilet paper roll in our bathroom about once every two weeks. At four hundred sheets per roll, that is thirty sheets per day for two people. As that is not the only toilet we use, 20 sheets per person per day is very reasonable.

At twenty sheets per person per day, we each use about 7×10^3 sheets per year. There are 3×10^8 Americans, so we use a total of

$$N = (7 \times 10^3 \frac{\text{sheets}}{\text{Am} \cdot \text{year}}) \times (3 \times 10^8 \text{ Am})$$

$$= 2 \times 10^{12} \frac{\text{sheets}}{\text{year}}.$$

That is 2 trillion sheets of toilet paper. Each sheet is about 4 inches or 10 cm long. Thus, the total length of

---

[*] Restricting ourselves to one sheet, as suggested by Sheryl Crow, might save paper [6] but would have other significant negative repercussions.

[†] If we get drunk and toss a few squares of toilet paper in the air, does that mean we are three sheets to the wind?

toilet paper used by all Americans is

$$L = \left(2 \frac{m}{Am \cdot day}\right) \times \left(400 \frac{days}{year}\right) \times (3 \times 10^8 \, Am)$$

$$= 2 \times 10^{11} \frac{m}{year} = 2 \times 10^8 \frac{km}{year}$$

That is about the distance from Earth to the Sun (although if we unroll it that far, the toilet paper will get very charred).

This is, of course, a very silly comparison. (It is also, of course, a very silly question.) A more reasonable comparison would be to our other daily paper usage. For example, our daily newspaper also contains about twenty sheets. However, each of those sheets has an area of about $1 \, m^2$, or about one hundred times larger than the $0.1 \, m \times 0.1 \, m = 10^{-2} \, m^2$ area of each sheet of toilet paper. If we want to reduce our environmental impact, we should start with the biggest items, not the smallest.

Money height

How tall is a stack of a trillion one-dollar bills?

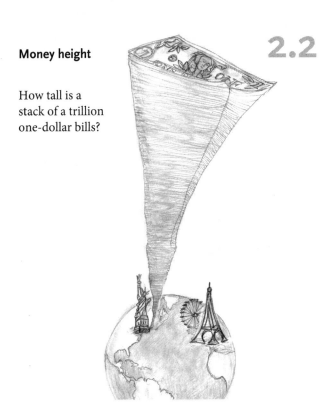

HINT: How thick is a ream (five hundred sheets) of paper?

ANSWER: To estimate this, we need to estimate the thickness of one bill. Perhaps you know the thickness of a single bill, but I don't. (Is it 0.1, 0.01, 10^{-3}, or 10^{-4} inches?)

When we can't easily estimate the thickness (or mass, or whatever) of a single object, we can often easily estimate the thickness of a set of objects. In this case, it is much easier to estimate the thickness of a ream (five hundred sheets) of paper than a single sheet. A ream of paper is about 2 inches (or 5 cm) thick. This means that the thickness of a single dollar bill is

$$t = \frac{5\,\text{cm}}{500\,\text{bills}} \times \frac{1\,\text{m}}{10^2\,\text{cm}} = 10^{-4}\,\text{m/bill},$$

and the height of a trillion (10^{12}) bills is

$$h = 10^{12}\,\text{bills} \times 10^{-4}\,\text{m/bill} = 10^8\,\text{m} = 10^5\,\text{km}.$$

Now we need to ask the crucial question. Is that big? Unfortunately, there is no natural scale, so let's choose some arbitrary ones. A distance of 10^5 km is about 6×10^4 mi (because 1 km = 0.6 mi). That is more than two times the circumference of the Earth and about 25% of the distance to the Moon.

Suppose that we could launch a trillion one-dollar bills into the atmosphere and keep them there. What fraction of the sunlight hitting the Earth could we block with all those dollar bills?

HINT: The radius of the Earth is $r = 6 \times 10^6$ m.
HINT: What is the area of the Earth?
HINT: What is the area of a dollar bill?

ANSWER: If we could somehow launch all of those dollar bills into the atmosphere, how much sunlight would they block? To estimate this, we need to estimate the area of all those dollar bills. Go ahead, take a bill out of your pocket and look at it. You can even splurge by looking at a $20 bill.

Each bill is about 2 inches by 6 inches, or 5 cm by 15 cm. This means that its area is 75 cm^2 or (rounding up) about 10^{-2} m^2. One trillion bills will then have an area of

$$A = 10^{12} \text{ bills} \times 10^{-2} \text{ m}^2/\text{bill} = 10^{10} \text{ m}^2.$$

And now, again, we ask: Is that big? We can compare this to the surface area of the Earth.* That is,

$$A_{\text{Earth}} = 4\pi r^2 = 4\pi (6 \times 10^6 \text{ m})^2 = 4 \times 10^{14} \text{ m}^2.$$

The fraction of the Earth's surface area that we could cover with 1 trillion one-dollar bills is

$$f = \frac{10^{10} \text{ m}^2}{10^{14} \text{ m}^2} = 10^{-4}$$

Thus, 1 trillion one-dollar bills would cover only one part in ten thousand of the Earth's area and therefore could block at most one part in ten thousand of the sunlight reaching the Earth. If we add the 2 trillion sheets of toilet paper used by Americans each year, we would only increase that to three parts in ten thousand.

That makes one trillion dollars seem tiny.

---

* Knowing a few numbers makes estimation easier (even in these days of searching the Web for everything). The radius of the Earth is 6×10^6 m = 6×10^3 km.

Really extra-large popcorn

How many kernels of popped popcorn are
needed to fill a bedroom? A movie theater?

ċ ċ

HINT: Estimate length, width, and height.
HINT: What is the volume of a room?
HINT: What is the volume of a piece of popcorn?

ANSWER: We need two pieces of information: the size of the average popped kernel of popcorn and the size of the average bedroom (or movie theater). A typical kernel of popcorn is about 1 cm to 2 cm in length and thus has a volume of about $5\,\text{cm}^3$—a value between $(1\,\text{cm})^3$ and $(2\,\text{cm})^3$. A typical bedroom is about 12 ft by 15 ft by 8 ft high, or 4 m by 5 m by 2.5 m. Thus it will have a volume of

$$V = (4\,\text{m}) \times (5\,\text{m}) \times (2.5\,\text{m}) = 50\,\text{m}^3.$$

Now we need to convert the units from cubic meters to cubic centimeters:

$$V = 50\,\text{m}^3 \times \left(\frac{100\,\text{cm}}{1\,\text{m}}\right)^3 = 50\,\text{m}^3 \times \left(10^6 \frac{\text{cm}^3}{\text{m}^3}\right)$$

$$= 5 \times 10^7\,\text{cm}^3.$$

This means that we will need

$$N = \frac{V_{room}}{V_{kernel}} = \frac{5 \times 10^7\,\text{cm}^3}{5\,\text{cm}^3} = 10^7$$

kernels of popcorn to fill the room. If we can count one piece per second, it will take four months to count them all.

A movie theater is significantly larger than a bedroom, although many people seem to use them interchangeably. A typical movie theater will have about twenty seats per row and about twenty rows. There will be about 3 ft (about 1 m) between seats and also 3 ft between rows. Thus the seating area is about $400\,\text{m}^2$ and the total area is about $600\,\text{m}^2$ (although this is getting needlessly precise). The average height is about 20 ft or 6 m, giving a volume of

$$V = (6\,\text{m}) \times (600\,\text{m}^2) = 4 \times 10^3\,\text{m}^3 = 4 \times 10^9\,\text{cm}^3.$$

Thus, the movie theater is about one hundred times larger than a bedroom and could contain about 10^9

kernels of popcorn (or about three kernels for every person in the United States).

Perhaps a more useful metric is the number of bags of microwave popcorn needed to fill a room. The volume of a full bag is about 1 liter, or about 10^3 cm^3. Thus, we would need to pop about

$$N = \frac{5 \times 10^7 \text{ cm}^3}{10^3 \text{ cm}^3/\text{bag}} = 5 \times 10^4 \text{ bags}$$

of popcorn to fill our friend's room. That would be expensive.

Building volume

What is the total volume of all human-built
structures (e.g., buildings, sheds, cars . . .)
in the United States? How does this
compare to the
volume of
Mt. Everest?

HINT: How large is your home?

HINT: How large is your workplace?

HINT: Mt. Everest is 10^4 m tall.

HINT: Mt. Everest is much broader than it is high.

ANSWER: There are lots of different kinds of structures: buildings, cars, airplanes, sheds, missile silos, boats. Of these, buildings clearly dominate. We can divide buildings into two general categories, places where we live and places where we work. Let's start with where we live. Most Americans live in houses. A typical house has an area of 1,000–2,000 ft^2 (100–200 m^2), a ceiling height of 8–10 ft (2–3 m), and houses two or three people. Taking the upper range of each estimate, this gives 600 m^3 for three people or 200 m^3/person. (If we took the lower range, we would get 200 m^3 for two people, or 100 m^3/person. As our goal is to get within a factor of ten of the correct answer, I refuse to worry about a factor of two.)

Now let's consider where we work or study. Most of us work in office buildings or attend school, where we each have a lot less space than 10^2 m^3/person. Many of us work in retail shops or restaurants where we also have significantly less room than 10^2 m^3/person. Very few of us work in factories (less than 10%) or farms (less than 2%), so this will not affect the results significantly. Let's use the higher estimate of 200 m^3/person in order to include work space as well as living space.

Thus, there are about 200 m^3 of human-built structure for each of the 3×10^8 Americans. This gives a total of

$$V_{build} = (200\,\text{m}^3/\text{Am}) \times (3 \times 10^8\,\text{Am})$$
$$= 6 \times 10^{10}\,\text{m}^3$$

or 60 km^3 of human-built structures in the United States.[*] This seems like a lot.

Now we need to compare this to something. We can compare it to something totally irrelevant (the

---

[*] We converted from cubic meters to cubic kilometers by multiplying by $(10^{-3}\,\text{km/m})^3 = 10^{-9}\,\text{km}^3/\text{m}^3$.

volume of a mountain) or to something slightly relevant (the area of the United States).

Let's start by comparing this to the volume of Mt. Everest. Because Mt. Everest is part of a mountain range and is not just an isolated peak, we will be making some highly questionable assumptions to estimate its volume. Mt. Everest is about 3×10^4 ft, or 10^4 m $= 10$ km, tall. We'll assume that it is a cone. The radius of the base will be between 10 km (giving an average slope of $45°$) and 100 km (giving an average grade of 10%) so we will use the geometric mean of 30 km. This gives a volume of

$$V = \frac{1}{3}\pi r^2 h = (30\,\text{km})^2\,(10\,\text{km}) = 10^4\,\text{km}^3.$$

Thus, the 60 km^3 volume of human-built structures in the United States is far less than the 10^4 km^3 volume of Mt. Everest.

Now let's make a (slightly) more relevant comparison. Let's calculate the average height our buildings would have if spread out over the entire area of the United States. The contiguous United States is about 3,000 miles (5×10^3 km) from east to west (a three–time zone or three-hour difference) and about 1,000 miles (1.5×10^3 km) from north to south, giving a total area of 7×10^6 km^2 $= 7 \times 10^{12}$ m^2. If all of our buildings are spread out evenly over this area, they would have an average height of

$$h = \frac{V_{build}}{A_{US}} = \frac{6 \times 10^{10}\,\text{m}^3}{7 \times 10^{12}\,\text{m}^2} = 10^{-2}\,\text{m} = 1\,\text{cm}.$$

That is definitely not a lot.

Mass of money

How much does $1 trillion weigh? Consider such media of exchange as dollar bills, large bills, gold, perfume, and printer ink.

HINT: What is the volume of 1 trillion one-dollar bills?

HINT: The density of paper is about 1 ton per m³.

HINT: What is the cost per ounce of perfume or gold?

HINT: What is the volume of printer ink in that $10 cartridge?

ANSWER: Let's start with one-dollar bills and move on to more esoteric media of exchange (ignoring the fact that most of these do not exist in trillion-dollar quantities). We already estimated the height of 1 trillion dollar bills to be 10^8 m and the area of one bill to be 10^{-2} m². Thus, the volume of 1 trillion dollar bills would be

$$V_{1\$} = hA = 10^8 \, \text{m} \times 10^{-2} \, \text{m}^2 = 10^6 \, \text{m}^3.$$

At a density of 1 ton per m³, this gives a mass of

$$M_{1\$} = (\text{density})(\text{volume}) = (1 \, \text{ton/m}^3)(10^6 \, \text{m}^3)$$

$$= 10^6 \, \text{tons}$$

or about the mass of ten nuclear-powered aircraft carriers.* Given that the largest denomination bill in circulation today is only $100, a trillion dollars in U.S. paper currency will have a minimum mass of 10^4 tons.† That is still way too much to carry easily. Note that as of June 2010, the total amount of all U.S. currency was only $0.9 trillion [7].

Now let's consider gold. Surely a trillion dollars in gold will be easier to handle. Gold currently costs about $1,000 per ounce.‡ This means that we need 10^9 ounces of gold. Because there are 16 ounces in a pound

---

* There are a plethora of different tons in use today, including short tons, long tons, and metric tons. They differ in mass by less than 10%, so we will ignore their differences and use 1 ton = 10^3 kg $\approx 2 \times 10^3$ lb.

† Larger bills were rendered obsoelete by electronic currency transfers.

‡ Yes, I know. Gold is measured in troy ounces, not in avoirdupois ounces. However, a troy ounce is only 10% heavier than an avoirdupois ounce. The larger difference between the two measures is that a troy pound contains only 12 troy ounces, whereas an avoirdupois pound contains the familiar 16 ounces. This means that an ounce of gold weighs more than an ounce of feathers but a pound of gold weighs less than a pound of feathers. Can we switch to the metric system now? Please?

and 2 pounds in a kilogram, there are 30 ounces in a kilogram. Therefore we need

$$M_{gold} = (10^9 \text{ oz}) \left(\frac{1 \text{ kg}}{30 \text{ oz}} \right) = 3 \times 10^7 \text{ kg}$$

$$= 3 \times 10^4 \text{ tons.}$$

Thus \$1 trillion of gold weighs less than 1 trillion one-dollar bills but more than 10 billion \$100 bills.

The gold will certainly occupy less volume than the currency. At a density of 20 tons per cubic meter, the 3×10^4 tons of gold "only" occupies a volume of $1.5 \times 10^3 \text{ m}^3$, which is just a bit smaller than the movie theater we filled with popcorn earlier.[*]

Now let's consider perfume. It costs between \$1 and \$1,000 per ounce (and probably higher). Let's take the geometric mean of \$30 per ounce. (OK, I admit it. I don't buy perfume. I cheated. According to the Internet, intermediately priced perfumes cost about \$20 to \$50 per ounce.) At 30 ounces per kilogram, that perfume costs $\$10^3$ per kilogram.[†] Thus, again we need 10^9 kg or 10^6 tons of perfume. Even if we used the most expensive perfume available, at a cost of \$3,000 per ounce, we would need 10^4 tons.[‡]

Now we get to ink for inkjet printers. It costs about \$5 to \$10 for a container of ink. A typical container is about 5 cm by 5 cm by 0.5 cm, or about 10 cm^3. That is about \$1 per cm^3 or \$1 per gram, so that ink for inkjet printers costs about the same as perfume.

[*] It will be much cheaper to fill the movie theater with popcorn than with gold.

[†] Yes, I am mixing volume ounces and weight ounces. However, because we are dealing with liquids that are mostly water and alcohol, the density is close enough to 1 that an ounce of perfume weighs about an ounce.

[‡] And if 10^4 tons of this perfume existed, it would be so common that it could no longer command such a high price, and we would need to buy something else.

Thus, $1 trillion of $100-bills, perfume, and inkjet printer ink would be the lightest, at a mere 10^4 tons. A trillion dollars of gold would be several times heavier but occupy about ten times less volume.

If you want to rob a bank, keep in mind that $1 trillion of $100-bills weighs 10^4 tons, $1 billion weighs 10 tons, and a mere $1 million weighs 10 kg. It's time to start pumping iron if you want to make crime really pay.

A baseball in a glass of beer

While I was at a Norfolk Tides baseball game, a foul ball landed in the section above me and showered some of my friends with beer. What is the probability of a foul ball landing in a cup of beer during one baseball game? What is the expected number of "splash downs" during all the major league baseball games played in an entire season? (See the answers for an even more improbable detail.)

HINT: How many foul balls land in the stands each game?

HINT: What is the size of a cup of beer?

HINT: What fraction of people have beer cups?

ANSWER: We need to break down the problem into manageable (or at least estimable) pieces. The first two pieces will be the number of foul balls per game that land in the stands and the probability that a given foul ball lands in a cup of beer. Let's start with the number of foul balls that land in the stands. The number per inning is definitely more than one and fewer than twenty, so we can take the geometric mean of five for our estimate.[*] If you watch a lot of baseball, you can probably come up with a better estimate; I counted several (between three and seven) foul balls per inning landing in the stands. With nine innings per game, this amounts to forty foul balls per game that could possibly land in a cup of beer.

Now we need to estimate the probability that a given foul ball will land directly in a cup of beer. (Note: only beer is sold in open-topped cups.) This means that we need to break the problem into even smaller pieces. Let's assume that the cup of beer is sitting innocently in a cup holder. To hit a cup of beer, the foul ball needs to

1. not be caught by a fan
2. land within the area of a seat
3. hit a seat whose owner has a cup of beer
4. land in the cup

Most fly balls are caught, but many are not. Let's estimate that between one-quarter and one-half of fly balls are not caught. "Averaging" the two, we will use one-third.

Most of the stadium area is used for seating, so let's ignore that factor.

At any given time, more than 1% and less than 100% of fans have a cup of beer in front of them.

[*] Yes, I know the square root of 20 is a touch less than 4.5. If you prefer to round down and use four rather than five, go ahead.

Using the geometric mean, we estimate that 10% of seats have beer cups.

A large beer cup is 4 inches (10 cm) across, so the baseball must land in an area defined by

$$A_{cup} = \pi r^2 = 3(2 \, \text{in})^2 = 10 \, \text{in}^2.$$

The area of the seat (from arm rest to arm rest and from row to row) is about 2 ft by 3 ft (60 cm by 90 cm), so

$$A_{seat} = (24 \, \text{in}) \times (36 \, \text{in}) = 10^3 \, \text{in}^2.$$

Thus, if the ball hits a seat that has a cup of beer, the probability that it lands in the cup is

$$P_{cup} = \frac{A_{cup}}{A_{seat}} = \frac{10 \, \text{in}^2}{10^3 \, \text{in}^2} = 10^{-2}$$

or 1%. The metric probability is the same.

(Extra credit question: Which is more likely, that the balls lands in the cup in the cup holder, splashing the beer, or that the fan is holding the cup of beer when the foul ball arrives and splashes it in his or her excitement?)

This means that the probability that any single foul ball lands in a cup of beer is

$$P = \frac{1}{3} \times \frac{1}{10} \times (10^{-2}) = 3 \times 10^{-4}.$$

With forty foul balls per game, this means that the probability of a foul landing in a cup of beer during any one game is 10^{-2}. This is not very likely. The probability that we will be directly below the splash is even less likely.

During the entire season, each of the 30 teams plays 160 games, giving a total of about 2,000 games (as it takes two teams to play a game). This means that the total number of beer landings in one season is

$$B = (2 \times 10^3 \text{ games per season})$$

$$\times (10^{-2} \text{ beer landings per game}) = 20.$$

Because baseball analysts keep meticulous statistics, I am very surprised that they do not appear to record beer landings.

Oh yes. The very improbable detail? According to my friends, the beer belonged to our former governor! ("Now at an improbability factor of a million to one against and falling," D. Adams [4].)

Life on the phone

How many people are talking or texting on their
cell phones at this instant?

HINT: What is the population of the world?

HINT: How much time do you spend each day talking on
your cell phone?

HINT: What fraction of time do you spend talking on your
cell phone?

ANSWER: The key insight to solve a problem like this is that the fraction of time the average person spends doing some activity is equal to the fraction of people doing it right now. For example, if the average person spends one-third of his or her time sleeping, then at this instant we can expect that one-third of all people are asleep. This means that

$$\frac{t_{talk}}{t_{day}} = \frac{N_{talk}}{N_{total}}$$

where t_{talk} is the average time spent talking or texting on the cell phone per day, t_{day} is the length of the day, N_{talk} is the number of people talking or texting on their cell phone right now, and N_{total} is the world population, which equals 7×10^9.

Now we need to estimate the amount of time the average person spends on his or her cell phone. We know that the proportion of cell phone use worldwide must be more than 1% and less than 100%, so we will take the geometric mean and estimate 10% cell phone use.[*] Similarly, each cell phone user will spend more than 1 minute and less than 10^3 minutes (16 hours!) talking each day, so we will take the geometric mean and estimate 30 minutes.[†] This means that average daily cell phone use, including everyone, is

$$t_{talk} = 30\,\text{min} \times 10\% = 3\,\text{min}$$

---

[*] These limits are broad enough to include the very high cell phone usage in the developed world and the much lower cell phone usage in the developing world.

[†] Your estimates may be less than mine. I work on a college campus, where it seems that almost all of the students are talking on their cell phones.

Thus

$$N_{talk} = \frac{N_{total} \, t_{talk}}{t_{day}}$$

$$= \frac{7 \times 10^9 \, \text{people} \times 3 \, \text{min/day}}{24 \times 60 \, \text{min/day}}$$

$$= 10^7 \, \text{people}$$

or 10 million people are talking on their cell phones right now. Wow! That is just as many as are picking their noses right now [8].

Despite the huge range between the upper and lower limits of our estimates, by taking the geometric means we can be reasonably confident that our answer is correct to within a factor of ten.

Money under the bridge

How much money
is collected per
hour during rush
hour at the George
Washington Bridge,
which connects
New York City
and New
Jersey?

HINT: How many lanes of traffic does the bridge carry?

HINT: How many cars per hour cross the bridge?

HINT: What is the toll for one car?

ANSWER: To estimate the money collected at the George Washington Bridge we need to estimate the traffic flow and the toll. In these days of E-ZPass and other electronic toll payers, it is easy to overlook the toll amount. However, the GWB toll passed $5 many years ago and is not yet at $20, so we will use the geometric mean of $10. Note that the toll is only collected heading into New York.[*]

There are two ways to estimate the traffic flow. We can estimate the number of toll booths and the time per car per toll booth. Alternatively, we can estimate the number of travel lanes and estimate the time per car per lane. Let's start with the second. The GWB has two levels, with about eight lanes per level. This means that there are eight inbound lanes. At the recommended time of 2 seconds between cars, each lane can handle 1,800 cars per hour for a total of 1.5×10^4 cars per hour.

Now let's estimate the traffic flow from the number of toll booths. The GWB has a *lot* of toll booths. There are definitely more than ten and fewer than one hundred, so let's take the geometric mean of thirty. The automatic toll booths can process one car every 3 to 4 seconds. The staffed toll booths take a lot longer. They can probably process one car every 10 seconds (fewer if you have exact change, more if you don't). About one-third of the toll booths are automated. This means that the automated booths can handle a total of three cars per second (ten booths divided by 3 seconds per car), and the staffed booths can handle a total of two cars per second (twenty booths divided by 10 seconds per car), for a total of five cars per second or 2×10^4 cars per hour. This is very close to the other estimate, lending credence to both.

---

[*] As it is unfair to make people pay to drive into New Jersey. N.B.: I grew up in New Jersey, so I know it's the landfill of opportunity.

Now we can calculate the money collected. At $10 per car, the GWB collects 2×10^5 per hour during rush hour. Wow. That's more than I collect in one entire year.

That is, of course, an irrelevant comparison. We should compare the money collected to the cost of the bridge. If we assume that the average collection rate is half of the peak rush hour rate, then the GWB collects a total of

$$T = \frac{1}{2} \times \$2 \times 10^5 \text{ per hr} \times 10^4 \text{ hr/yr} = \$10^9 \text{ per year}$$

A new bridge would cost more than $1 billion and less than $100 billion, so we'll estimate $10 billion. My mortgage payments are about 1% of the cost of my house per month, or about 10% per year. Applying the same logic to the George Washington Bridge, the "mortgage" payments would be 10% of $10 billion, or about 10^9 per year. This seems reasonable.

Thanks to Chuck Adler of St. Mary's University in Maryland for the question.

If a million monkeys type randomly on a million typewriters for a year, what is the longest string of consecutive correct letters of Shakespeare's *Hamlet* (or *The Cat in the Hat* [9]) (starting from the beginning of the play) that they would likely type (assuming that they typed only letters and ignoring capitalization, punctuation, and spacing)? If all of the world's computers generate random letters for one year, what is the longest string of consecutive correct letters of Shakespeare's *Hamlet* that they would likely generate?

HINT: The chance that the first letter is correct is 1 out of 26.

HINT: The chance that the first two letters are correct is 1 out of 26^2 or about 1 in 1,000.

HINT: How many random characters per second can a monkey type?

HINT: There are $\pi \times 10^7$ seconds in a year.

HINT: How many computers are there in the world?

HINT: How many random characters can one computer generate per second?

ANSWER: We need to estimate the total number of letters typed and the probability of finding a string of n correct consecutive letters in that total. An expert typist can type one hundred words per minute correctly. At about five letters per word, this works out to about ten letters per second. Thus, we'll assume that any typist can type ten letters per second randomly.[*] In one year the million monkeys will type

$$N = 10^6 \text{ monkeys} \times (\pi \times 10^7 \text{ s}) \times 10 \text{ letters/s}$$

$$= 3 \times 10^{14} \text{ letters}$$

If we insist on letting them eat, drink, sleep, etc., then they will only type $N = 10^{14}$ letters.[†]

The probability that the first letter is correct is 1 out of 26. The probability that the second letter is correct is 1 out of 26. The probability that the nth letter is correct is also 1 out of 26. This means that the probability that n consecutive letters are correct is 1 out of 26 raised to the nth power, or 26^n. If the probability of an event happening—say of a coin landing on its edge—is 1/10,000, then we would have to toss about 10,000 coins for one to probably land on edge. So if the probability of getting a particular string of n letters is $1/26^n$, then we need about $N = 26^n$ trials to probably get such a string.

For example, out of a string of twenty-six random letters, we would expect that one would match the first letter of *Hamlet* (or *The Cat in the Hat*). Out of a string of $26^2 = 676 \approx 10^3$ letters, we would expect that one set of two would match the first two letters of *Hamlet* and out of a string of 10^6 letters, we would expect that about one set of four would match the

[*] And yes, we will ignore all of the tremendous difficulties inherent in typing truly randomly. Random number generation is a sophisticated and subtle field.

[†] Remember that 1 year $= \pi \times 10^7$ s, a very useful conversion factor.

first four letters of *Hamlet*. Therefore, for every factor of 10^3 increase in the number of random letters, the length of the longest correct string of letters should increase by two.[*]

Let's splurge and use 10 million monkeys. Then we will have a string of 10^{15} random letters. At two correct letters for every factor of 10^3 random ones, we can expect ten correct consecutive letters. That is not a lot.[†]

Now let's consider the computers. We need to estimate the number of computers in the world and their random letter generation speed. There are more than 10^8 and fewer than 10^{10} computers in the world (as the developed world has at least one computer per household and the entire world has fewer than one computer per person). We will take the geometric mean and use 10^9 computers.[‡] The average computer runs at about 1 GHz (more than 100 MHz and less than 10 GHz) and requires ten instructions to generate a random letter (more than one and fewer than one hundred) and can thus generate 10^8 random letters per second. This means that in one year the computers will generate

$$N = (10^9 \text{ computers}) \times \left(10^8 \frac{\text{letters}}{\text{comp-s}}\right)$$
$$\times (\pi \times 10^7 \text{ s}) = 3 \times 10^{24} \text{ letters.}$$

Wow. That is five moles of letters!

---

[*] To be precise (which is an epithet in this book), if there are N random letters, then the expected length of the longest correct consecutive string of letters is
$n = \log_{26} N = \log N / \log 26 = \log N / 1.4$. (While the logarithm might appear daunting, $\log N$ is just the exponent of N.) This is within 10% of our rough calculation.

[†] And by the way, good luck finding that string of ten correct letters out of the quadrillion random ones!

[‡] Even if our estimate of the number of computers is off by 10^3, that will only change the length of correct text by two letters.

But at two correct letters for every three orders of magnitude, that is only sixteen correct letters (or only six more than those 10 million monkeys produced). If we ignore stage directions, we are left with:

ACT I

Scene I. Elsinore. A platform before the castle.
Francisco at his post. Enter to him Bernardo

BERNARDO: **Who's there?**
FRANCISCO: **Nay, answ**[*]

Not too impressive.

---

[*] Or perhaps we prefer: "The sun did not shin". That's probably in there too.

The titans of siren

How many ambulances would be needed in the United States so that everyone can be reached within 8 minutes in an emergency?

HINT: How far can an ambulance travel in 8 minutes?

HINT: How much area can it cover in that time?

HINT: What is the area of the United States?

HINT: It is about 3,000 miles from coast to coast.

ANSWER: In order to estimate the number of ambulances needed to cover the United States, we need to estimate the area that one ambulance can cover in 8 minutes and then divide that into the area of the United States.[*] It will take 1 to 2 minutes for the ambulance crew to get in the ambulance. However, as this is much less than 8 minutes, we will ignore that. An ambulance can probably travel about 30 mph on city streets and 50 mph on country roads.[†] Because most of the U.S. land area is not urban (although most of the population is urban), let's use 50 mph. This means that an ambulance can travel a distance

$$r = (50 \,\text{mi/hr}) \times (8 \,\text{min}) \times \left(\frac{1 \,\text{hr}}{60 \,\text{min}} \right)$$

$$= 7 \,\text{mi}$$

and the area covered by one ambulance is

$$A = \pi r^2 = 150 \,\text{mi}^2$$

Now we need the area of the United States. You might remember that it is about 3,000 miles from east to west or that it is about 1,000 miles from north to south. On the other hand, you might remember that it is a difference of three time zones from east to west, which is $3/24 = 1/8$ the circumference of the globe. On the gripping hand, you might remember that it is a 6-hour plane flight and that planes fly at about 500 mph. However you remember it, this would let us estimate the area of the United States as

$$A = (3 \times 10^3 \,\text{mi}) \times (10^3 \,\text{mi}) = 3 \times 10^6 \,\text{mi}^2$$

[*] This is the response time for an advanced life support ambulance according to National Fire Protection Association standards [10].

[†] Unlike most of the problems in this book, here it is equally easy to work in either U.S. customary or metric units. Because we are just dividing two areas, there are none of the complicated unit conversions that make U.S. customary units so hard to use.

The number of ambulances required is just the ratio of the two numbers:

$$N = \frac{3 \times 10^6 \, \text{mi}^2}{150 \, \text{mi}^2}$$
$$= 2 \times 10^4$$

According to the American Ambulance Association [11], there were more than forty-eight thousand ambulances in the United States in 2004. Given that the need for ambulances depends on population density and demographics in addition to just coverage area, this is a quite reasonable estimate.

Airheads at the movies

Which has more mass, the air or the brains in a movie theater?

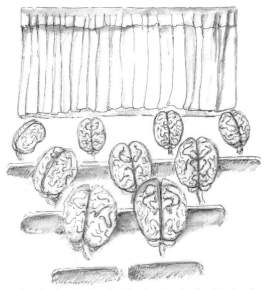

HINT: What is the volume of a movie theater? Consider its height, width, and depth.

HINT: Air has a density of about 10^3 times less than water.

HINT: Water has a densiry of 10^3 kg/m^3.

HINT: What is the size of the brain in liters?

HINT: A liter of water has a mass of 1 kg.

HINT: How many people can sit in a movie theater?

ANSWER: In order to compare these, we need to estimate the volume and density of both the air and the brains in a typical movie theater. We estimated the volume of a movie theater earlier in this chapter (when we wanted to fill it with popcorn) as

$$V = 4 \times 10^3 \, \text{m}^3.$$

When liquids turn to gas, they expand by a factor of about one thousand.[*] This means that the density of air is about 10^3 times less than the density of water,[†] $\rho_{water} = 10^3 \, \text{kg/m}^3$, so that $\rho_{air} = 1 \, \text{kg/m}^3$. Therefore the mass of all that air is

$$M = \rho_{air} V = (1 \, \text{kg/m}^3) \times (4 \times 10^3 \, \text{m}^3)$$

$$= 4 \times 10^3 \, \text{kg},$$

or about 4 tons.

Assuming that the movie theater is fully occupied, and neglecting the brains of the actors,[‡] there are four hundred brains in the room. The volume of a human head is more than 1 liter and less than 4 liters (1 gallon), so we can take the geometric mean and use 2 liters. About half of that volume is brain, at a density of about water. Thus, the total mass of all the brains in the movie theater is

$$M = \rho_{water} V = (1 \, \text{kg/L}) \times (4 \times 10^2 \, \text{L})$$

$$= 4 \times 10^2 \, \text{kg}.$$

Therefore, in a typical movie theater, there are 400 kg of brains and 4,000 kg of air.

Feel free to draw your own conclusions.

---

[*] This is another useful fact worth remembering or inscribing on your shirt cuff.

[†] The density of water is $\rho_{water} = 1 \, \text{g/cm}^3 = 1 \, \text{kg/L} = 10^3 \, \text{kg/m}^3$. Different units are easier for different problems.

[‡] Which would be reasonable even if they were physically present.

Heavy cars and heavier people

At the fiftieth anniversary celebration for the Golden Gate Bridge, people thronged the roadway. So many people crowded onto the bridge that its arch flattened, and worried engineers checked the load limits. Compare the mass density (in mass per area) of cars on the bridge on a normal day and people on the bridge during the celebration.

HINT: What is the mass of a car?

HINT: How far apart are cars on the bridge under normal conditions? How much area does a car "occupy"?

HINT: What is the mass of a person?

HINT: How far apart are the people on the bridge? How much area does a person occupy?

ANSWER: Cars are much heavier than people, but they occupy significantly more space. Let's estimate the space occupied by one car in both freely flowing traffic and a traffic jam.

In both cases, the lane width will be 12 ft (or about 4 m). In a traffic jam, there will be about one car length between cars. Given that a typical car is about 12 ft (4 m) long, this means that a car in a traffic jam occupies an area of 12 ft by 24 ft, or about 300 square ft (about 30 m^2). A typical car has a mass of 1 to 2 tons. Using 2 tons, this gives a mass per area of

$$\left[\frac{M}{A}\right]_{jam} = \frac{2,000 \text{ kg}}{30 \text{ m}^2} = 60 \text{ kg/m}^2.$$

In freely flowing traffic, drivers should allow 2 seconds of following distance. This means that at 60 mph (about 30 m/s), drivers should leave 60 m between cars. Thus, at highway speeds each car should occupy an area of $A = 4 \text{ m} \times 60 \text{ m} = 240 \text{ m}^2$ (or about ten times as much area). Thus the mass density will be almost ten times smaller, or

$$\left[\frac{M}{A}\right]_{60mph} = \frac{2,000 \text{ kg}}{2 \times 10^2 \text{ m}^2} = 10 \text{ kg/m}^2.$$

Now the task of estimating the human density should be rather pedestrian. How many people crowd into a square meter on festive occasions? One person per square meter would provide plenty of elbow room and five people per square meter would be approaching Tokyo subway densities. Let's estimate two people per square meter. At 100 kg per person (in round numbers), that gives a human density of

$$\left[\frac{M}{A}\right]_{people} = 200 \text{ kg/m}^2$$

That is thirty times the mass density of free-flowing traffic and three times the mass density even of a traffic jam in both directions. Who knew that people were so heavy?

The engineers were quite right to worry.

Thanks to Ralph E. Edwards of Norfolk, Virginia, for suggesting the question.

Peeing in the pool

How much urine
is in a typical public
swimming pool?

‿‿‿‿‿‿‿‿‿‿‿‿‿‿‿‿‿‿‿‿‿‿‿‿

HINT: How many children use the pool every day?

HINT: What fraction of the children pee in the pool?

HINT: What is the volume of urine excreted per child?

HINT: How rapidly is the urine removed from the pool?

ANSWER: A good friend told me that his summer swim team had been accused by the recreational swim director of peeing in the community swimming pool. In response, they quietly dumped a gallon of phenolphthalein (a chemical that turns bright red in the presence of acid) into the pool. During the following children's swim period red regions periodically bloomed around some of the children. The accusations ceased, and the water stayed pink for a week. Unfortunately, according to Snopes [12], there is no chemical that will reveal the presence of urine in a swimming pool. However, there is almost certainly some urine in our local public swimming pool.

In order to estimate the amount of urine in the pool, we just need to estimate the rate at which urine enters the pool and the rate at which it leaves the pool. The rate of urine entering the pool depends on the number of children using the pool, the fraction of children peeing in the pool, and the amount of urine excreted.

Let's look at the worst case, a popular public outdoor pool in the summer. As usual, we will determine upper and lower bounds for our quantities. The pool will have about fifty children in it from morning to evening (more than ten and fewer than two hundred). About 30% of the children will pee in the pool (more than 10% and less than 100%). Between 9 a.m. and 9 p.m. each child who pees in the pool will do so twice.

Now we need to estimate a typical urination volume. There are several approaches. We can try to (1) estimate the volume directly or (2) estimate the daily volume of liquid consumed and divide by the number of times we pee (assuming no perspiration) or (3) estimate the volume flow rate and the time.

My job does not require me to pee into a cup, so let's skip method 1. I consume about 16 oz (0.5 liter) of liquid each for breakfast, coffee, lunch, and dinner, for a total daily consumption of 2 L. I probably pee about five times a day, giving a volume per micturition of 0.4 L. Now let's consider method 3. Males have an advantage here, as we get to watch the outflow. The flow stream is a few millimeters in diameter, and the flow velocity is probably 1 to 2 meters per second.[*] This gives a volume flow rate of $(0.3\,\text{cm})^2(200\,\text{cm/s}) = 20\,\text{cm}^3/\text{s} = 20\,\text{mL/s}$. If it takes about 10 seconds to pee (more than 1 second and less than 100 seconds), this gives a volume of 200 mL, or 0.2 L. Let's go with the input-based method and use 0.4 L for an adult. A child will have a volume of about half of that, so we'll use 0.2 L.

This means that urine will enter the pool at a rate:

$$V_{pee} = 50\,\text{children} \times 0.3 \times 2\,\frac{\text{urinations}}{\text{day}}$$

$$\times 0.2\,\frac{\text{L}}{\text{urination}} = 6\,\frac{\text{L}}{\text{day}}.$$

This seems like a lot, but pools are big. To estimate their volume, we need to estimate the length, width, and depth. A typical pool will be 25 m (or yards) long by six lanes wide. Each lane is about 2 m wide (so that two or more swimmers can swim in each lane). The pool will vary in depth from 1.5 m (5 ft) in the shallow end to about 4 m in the diving end, with an average depth of 3 m. This gives a pool volume of

$$V_{pool} = 25\,\text{m} \times 12\,\text{m} \times 3\,\text{m} = 10^3\,\text{m}^3 = 10^6\,\text{L}.$$

Thus, urine is added to the pool at a rate of 6 ppm (parts per million) per day.

---

[*] If we want to determine the flow velocity more precisely, measure how far away the flow strikes the ground when the initial velocity is horizontal. It takes slightly less than 0.5 s for the liquid to fall 1 m, so if the flow strikes the ground 1 m from its starting point, then its velocity is slightly more than 2 m/s.

Now we need to estimate the rate at which urine is removed from the pool. This depends on the water purification methods. It is rather complicated and depends on many factors.[*] However, the important quantity is the time period in which the urine is removed. The removal period will be more than one day and less than one hundred days. Let's take the geometric mean and estimate that urine remains in the pool for ten days.

This means that there is, on average, 60 L of urine in a public outdoor swimming pool and that this amounts to 60 parts per million. Fortunately, urine is sterile.

---

[*] Urine is also a complicated collection of different molecules with different removal rates. We will ignore that complication.

Recycling: What Really Matters?

Chapter 3

We are constantly urged to recycle in order to "save the Earth." How much effort should we really put into recycling? Given our limited resources (including especially mental effort and available attention), where should we put our environmental efforts?

Note that there are two schools of thought about environmental and other moral efforts. The first school claims that if we can just train everybody to be environmentally conscious, then everybody will recycle, drive smaller cars, bicycle to work, and generate their own electricity during gym workouts. The second school claims that humans perform only enough "good deeds" to see themselves as good people [13, 14]. The concept of "moral licensing" explains such seemingly anomalous behavior as driving a 10-mpg SUV around the block to go to an organic food store, riding the elevator at the gym, and leaving high-efficiency lights on longer than regular ones [15]. If the moral licensing school is correct, then we should carefully select which environmental efforts we perform.

Remember that there is plenty of room in the United States for landfills. As we estimated in *Guesstimation*, a 100 m-tall landfill that covered 10^3 km^2, or 400 mi^2, could hold *all* of our garbage (at 2×10^8 tons per year) for an entire century. That's the area of a city like Virginia Beach or Los Angeles or one part in ten thousand of the area of the United States. Thus we recycle to save resources, not to save landfill space.

Typically, the most important resource is energy. While there are several different kinds of energy, the standard (metric) unit for all of them is the joule (although, as usual, we will need to convert among several different units).

Glassmaking uses thermal energy to convert sand (plus a smattering of carefully chosen contaminants) into glass. The important quantity is the specific heat capacity of the material, which is defined as the amount of energy required to raise the temperature of a kilogram (or gram or pound or . . .) of the material by 1 degree. The specific heat capacity of water is

$$C_{water} = \frac{heat}{mass \times temperature\ change} = 1\frac{cal}{g \cdot {}^\circ C}$$

$$= 4 \times 10^3 \frac{J}{kg \cdot {}^\circ C}.$$

Other materials will have specific heat capacities that are much lower than water. Note that a calorie (1 cal = 4 J) is defined as the energy needed to raise 1 gram of water by 1 degree celsius.[*]

Power measures how fast we use energy and is defined as the energy used divided by the time. A watt (W) is a joule per second (1 W = 1 J/s). A 100-W light bulb uses 100 J every second. Electric companies typically charge for their energy in units of kilowatt-hours (kWh), which is 1 kilowatt of energy used for 1 hour. So,

$$1\,kWh = (10^3\,W) \times (4 \times 10^3\,s) = 4 \times 10^6\,J,$$

where 1 hour equals $60 \times 60 = 4 \times 10^3$ seconds. One kilowatt-hour costs about $0.10 in most of the United States.

Chemical reactions rearrange chemical bonds and transform chemical energy to or from other forms of energy. The typical energy of a strong chemical bond is 1.5 eV. We know this because common batteries, which convert chemical energy into electrical energy, operate at 1.5 V. It takes one chemical reaction to move one electron through a potential difference of 1.5 V, changing its energy by 1.5 electron volts (eV).

---

[*] A British thermal unit (BTU) is similarly defined as the energy needed to raise 1 pound of water by 1 degree Fahrenheit.

Because there are 6×10^{18} electrons in a coulomb, there are 6×10^{18} electron volts in a joule (1 J = 1 coulomb volt).

Gasoline and other hydrocarbons contain chemical energy. When the carbon or the hydrogen combine with oxygen (C + O_2 → CO_2 or H_2 + O → H_2O), 1.5 eV of energy is transformed into heat. For estimation purposes there are at most three hydrocarbons: carbon, CH_2, and CH_4. When we convert from eV per reaction to joules per kilogram, we find that one kilogram of carbon (e.g., coal) contains 2×10^7 J; 1 kilogram of CH_2 (e.g., gasoline) contains 4×10^7 J; and one kilogram of CH_4 (e.g., natural gas or methane) contains 6×10^7 J of chemical energy.[*]

We will discuss other forms of energy, such as energy of motion and energy of position, in a later chapter.

[*] See *Guesstimation* for details.

Water bottles

A visiting colleague once spent about 10 minutes trying to figure what to do with his empty plastic water bottle. This seemed like an inordinate amount of time to expend on such an insignificant item and prompted this question: If there is a trash can at our elbow, how much further should we walk to recycle a plastic water bottle?

ċ ċ

HINT: What is the energy content of the plastic?
HINT: Hydrocarbons have an energy content of about 4×10^7 J/kg.

ANSWER: Because plastics contain far more complicated molecules than do aluminum, steel, or glass (SiO_2), recycling plastics is far more complicated than recycling metal and glass. The plastic in plastic water bottles is usually recycled to make plastic fiber rather than more water bottles.

We can set a lower limit on the value of recycling a plastic bottle by estimating the energy content of the plastic (i.e., the energy that would be released by incinerating it). In order to do this, we need to estimate its mass and energy density (the energy per kilogram).

Let's consider a 16-oz plastic water bottle. The bottle itself weighs far less than its contents. Although it is difficult to estimate the mass of very light large objects, we can establish upper and lower bounds. A plastic water bottle definitely weighs more than 1 gram (the mass of a paperclip) and less than 100 g (the mass of 4 ounces or $\frac{1}{2}$ cup of water). Taking the geometric mean, we will estimate its mass at 10 g, or 10^{-2} kg.

Now we need to estimate the energy density of plastic. Because plastic is a hydrocarbon, its energy density should be similar to that of gasoline. This means that the energy content of one bottle should be

$$E = (10^{-2} \text{ kg})(4 \times 10^7 \text{ J/kg}) = 4 \times 10^4 \text{ J}.$$

This seems like a lot. However, as usual, we need to compare it to something relevant. Electrical energy is typically measured in kilowatt-hours (kWh). Given that 1 kWh $= 4 \times 10^6$ J, the energy content of a plastic bottle is 10^{-2} kWh. At \$0.10 per kWh, this is about one-tenth of a penny.

This is just a lower limit. The real value of the recycled bottles is to replace the crude oil (and its distillates) needed to make new plastic. The amount of oil needed to make a plastic bottle will be more than one and less than a hundred times the bottle's mass. Taking the geometric mean, we should multiply the energy value of the bottle estimated above by a factor of ten. This means that the value of a recycled bottle is an entire penny.[*]

We can check this estimate. At a penny per 10-g bottle, this implies a wholesale price for sorted, shredded plastic of $1 per kg, or $1,000 per ton. As of July 12, 2010, the recycled resin price for clear PET bottles was about $0.50 per pound, or about $1 per kg [16].

Note that we are ignoring the cost of transporting that bottle to the landfill. The bottle would need to be transported anyway, whether to the landfill or the recycling center. The cost of the landfill itself is negligible, as we only need a tiny part of the United States (10^{-4}) to landfill all of our trash for an entire century.

So how far should we be willing to walk to recycle that water bottle? Let's assume that we are selfless and willing to put in one penny's worth of our own effort in exchange for one penny's worth of benefit to the environment. We need to estimate the value of our time. The most convenient[†] measure of the value of our time is the amount that our employers are willing to pay us for it. The U.S. per capita gross domestic product (as of 2008) was $40,000 [17]. This understates the average wage, because not all Americans work for pay, yet it also overstates the average wage, because it includes non-wage income. Fortunately, for purposes of estimation, we don't care (as it is probably within a factor of two and certainly within a factor of ten of the correct value).

Assuming 40 hours per week and fifty weeks per year, that $40,000 per year translates to an hourly wage

[*] Although we are ignoring the cost to separate it, sort it, wash it, and shred it.

[†] Although not necessarily the most accurate.

of \$20. At about 4,000 seconds per hour, 2 seconds of our time is worth one penny (unless, of course, it is time spent reading this book, in which case it is far more valuable). At a normal walking pace, we can walk about 3 m (10 ft) in 2 seconds. That is not very far.

There's another comparison we can make. If we use one plastic water bottle per day, in one year that will amount to a mass of

$$M = (10^{-2}\,\text{kg/bottle}) \times (4 \times 10^2\,\text{bottles}) = 4\,\text{kg}$$

of plastic. If we drive the average amount of 10^4 miles per year at 20 miles per gallon, we consume 500 gallons or 2×10^3 kg (2 tons) of gasoline per year. The oil needed to make those four hundred water bottles is negligible by comparison.

Now is it worth spending 10 whole minutes figuring out how to recycle that bottle?

How far should you be willing to walk
to recycle a glass beer bottle?

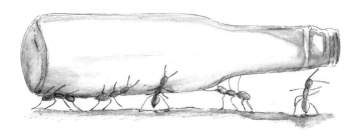

꜡ ꜡

HINT: How much energy does it take to make a glass
bottle?

HINT: Glass is made from sand.

HINT: The heat capacity of water is 1 cal/g · °C or
4×10^3 J/kg·°C.

HINT: What is the cost of that energy?

ANSWER: In the United States, recycled glass bottles are crushed and used as raw material to make new bottles. The value of a used glass bottle is then, at most, the cost of its production. That includes the value of its raw materials and of the energy needed to transform those raw materials into glass. The raw material is primarily silicon dioxide, otherwise known as sand.* As this is the most common material in the Earth's crust [18], we will neglect its cost. The sand is transformed into glass by heating it to about 1,500 °C. In order to estimate the energy needed to make one bottle's worth of glass, we need to estimate the mass of the bottle and the heat capacity of sand.†

I am not very good at estimating the mass of an individual object, so let's estimate how many bottles weigh a pound (or mass a kilogram). There are more than two and fewer than six bottles per pound (because an empty six-pack weighs more than a pound), so let's take the average of four bottles per pound (or ten bottles per kilogram).‡

The heat capacity of water is 4×10^3 joules (or 10^3 calories) per kg per degree Celsius. This is one of the highest heat capacities of any common material. Let's use a heat capacity for SiO_2 that is a lot less than that of water and is also a nice round number: $c = 10^3$ J/kg · °C.§

---

* Yes, I know this is imprecise nomenclature. "Sand" refers to the size of the grain, rather than to the material. However, most sand is quartz sand and is primarily composed of silicon dioxide.

† The heat capacity is the amount of energy needed to raise the temperature of 1 kilogram of the material by 1 degree Celsius.

‡ I confess with shame that I cheated. After making this estimate, I went home and weighed a beer bottle. (In the interests of science, I emptied it first.) The bottle weighed 7 oz, or 0.2 kg. Although this is within the upper and lower bounds, it is twice as large as our estimated mass of 0.1 kg.

§ The actual heat capacity of SiO_2 is 7×10^2 J/kg·°C.

Thus, in order to heat one bottle up to 1,500 °C, it requires energy

$$E = cm\Delta T = (10^3 \text{ J/kg} \cdot {}^{\circ}\text{C})(0.1 \text{ kg})(1,500 \, {}^{\circ}\text{C})$$
$$= 1.5 \times 10^5 \text{ J}.$$

This seems like a lot. However, as usual, we need to compare it to something relevant. Since 1 kWh (4×10^6 J) of electricity costs about $0.10, the energy required to turn 0.1 kg of sand into 0.1 kg of glass is less than one penny.

We have a recycling bin next to our garbage bin at home. The effort needed to put a bottle in the recycling bin is the same as the effort needed to put it in the trash bin. However, when we are away from home, it is just not worth trying to find a recycling bin for a few pennies worth of glass.

How much effort are you willing to invest in order to save a penny?

How far should you be willing to walk to recycle
an aluminum can?

ċ ċ

HINT: Each aluminum atom in bauxite (aluminum ore) is
tightly bound.
HINT: A typical chemical reaction in, e.g., a battery, is 1.5
electron volts per atom.

ANSWER: The value of the aluminum is equal to the cost of the raw material plus the energy needed to process it. Aluminum is refined from bauxite ore. Each aluminum atom is very tightly bound and much harder to purify than, say, from iron ore. Energy is supplied during the smelting process to free those atoms. In order to estimate the energy needed, we need to estimate the energy of a typical chemical reaction and the number of reactions (i.e., the number of atoms involved).

Fortunately, we have lots of chemical reactors close at hand. The technical term for these reactors is "batteries." Common batteries convert chemical energy to electrical energy. Each electron acquires an electrical potential of 1.5 V, so each electron gains an energy equal to 1.5 electron volts. Thus, we estimate that it takes 1.5 eV to remove each aluminum atom from bauxite.[*]

Now we just need to estimate the number of aluminum atoms. The atomic weight of aluminum is 27 (more than $A = 1$ for hydrogen and less than $A = 200$ for uranium), so 1 mole of aluminum contains $N_A = 6 \times 10^{23}$ atoms and has a mass of 27 g. Thus, one kilogram of aluminum contains, $\frac{1000}{27} = 40$ moles or 2×10^{25} atoms. This means that the energy needed to purify 1 kg of aluminum is

$$E = \frac{(1.5 \text{ eV/atom})(2 \times 10^{25} \text{ atoms/kg})}{(6 \times 10^{18} \text{ eV/J})}$$
$$= 5 \times 10^6 \text{ J/kg}.$$

We should probably double that to account for inefficiencies in the processing.

Let's check with reality before proceeding. The heat of reaction for reducing aluminum oxide (Al_2O_3) is about 30 MJ/kg, or six times greater than our estimate.

[*] It also really helps to know the conversion from eV to joules:
$1 \text{ J} = 6 \times 10^{18} \text{ eV}.$

The actual electrical energy required to produce aluminum is about 50 MJ/kg. Our mistake was to use the typical chemical reaction energy of a battery and not account for the fact that aluminum oxide is much more tightly bound.[*] (It is so tightly bound that it has to be reduced electrically rather than chemically.)

Now let's consider that soda can. A soda can weighs about the same as a plastic water bottle (give or take a factor of two), so we'll use the same estimate of $m = 10$ g $= 10^{-2}$ kg. Thus, the electrical energy needed to produce the aluminum for our soda can is

$$E_{can} = (5 \times 10^7 \text{ J/kg})(10^{-2} \text{ kg}) = 5 \times 10^5 \text{ J}.$$

That is also about one-tenth of a kilowatt-hour, or about one penny. Even if we multiply this number by four to account for nonelectrical costs, it's still only a few pennies.

[*] I am willing to take the blame this time.

If the number of times we are asked a question is related to its importance, then "paper or plastic?" is the single most important environmental question of our time. What is the total mass of all of the grocery store plastic bags used by an average American in one year? How does this compare with the other hydrocarbon products we use?

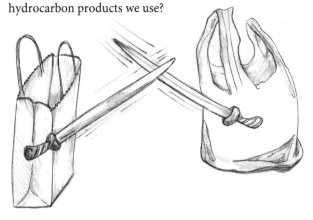

¿¿¿¿¿¿¿¿¿¿¿¿¿¿¿¿¿¿¿¿¿¿¿¿¿

HINT: How much does a roll of plastic bags weigh?

HINT: How many grocery store plastic bags do you use each week?

ANSWER: In order to estimate the mass of all the plastic bags we use each year, we need to estimate the number of bags we use and the mass of each bag.

The average American family of four[*] shops twice per week[†] and buys ten plastic bags of groceries each time. This means that they use

$$n = (2 \text{ trips/wk}) \times (50 \text{ wks/yr}) \times (10 \text{ bags/trip})$$

$$= 10^3 \text{ bags/yr},$$

or about three hundred bags per person per year. Therefore the 3×10^8 Americans will use

$$N = (300 \text{ bags/person} \cdot \text{yr}) \times (3 \times 10^8 \text{ persons})$$

$$= 10^{11} \text{ bags/year}.$$

That is a huge number of bags. Before we jump to conclusions, let's estimate their mass.

There are different ways to estimate the mass of the average grocery store plastic bag. One bag seems to have about the mass of one paper clip, or about 1 gram; however, it is very difficult to compare the masses of such disparate items. The stack of a hundred or so bags in the grocery store has a mass of a few ounces (more than 1 ounce and less than 1 pound). At about 30 g per ounce, this also gives a mass per bag of about 1 g.

We can also estimate the mass from the product of the volume and the density. One bag is about 30 cm × 30 cm (or 1 ft × 1 ft). Estimating the thickness is more difficult. Those one hundred bags in the grocery store are about 1 cm thick (definitely more than 1 mm and less than 10 cm), so each bag is about 10^{-2} m / 100 = 10^{-4} m thick. We can compare this to the thickness of aluminum foil ($\approx 25 \times 10^{-6}$ m) or plastic wrap ($\approx 10 \times 10^{-6}$ m). This gives a volume per bag of

---

[*] Making the very dubious assumption that my family is average.

[†] Less than that if they are organized and more than that if they are disorganized.

$V = (0.3\,\text{m})^2(10^{-4}\,\text{m}) = 10^{-5}\,\text{m}^3$. As plastic has about the same density as water ($\rho = 1\,\text{g/cm}^3 = 10^3$ kg/m^3), this gives a mass $m = 10^{-5}\,\text{m}^3 \times 10^3\,\text{kg/m}^3 = 10^{-2}$ kg = 10 g. Let's take the geometric mean of the different estimates and use $m = 3$ g.

The mass of an individual American's bags is

$$m_{\text{bags}} = \left(300\,\frac{\text{bags}}{\text{year}}\right)\left(3\,\frac{\text{g}}{\text{bag}}\right) = 1\,\frac{\text{kg}}{\text{year}}.$$

Before we know whether this is a little or a lot, we need to compare it to something else. Because plastic bags are made from hydrocarbons, let's compare this to the mass of hydrocarbons we burn in our cars.

The average American drives about 10^4 mi/yr (using the average mileage used in car warranties).[*] Our car or SUV gets about 20 mi/gal,[†] so it burns about

$$V_{\text{gas}} = \frac{10^4\,\text{mi/yr}}{20\,\text{mi/gal}} = 500\,\text{gal/yr} = 2 \times 10^3\,\text{L/yr}.$$

Given that gasoline has about the density of water, each car burns about 2×10^3 kg per year of gasoline. That is 2 *tons* per year.

This means that average Americans use thousands of times less hydrocarbons in their grocery store plastic bags than in their automobiles. Carefully keeping our tires fully inflated will save far more hydrocarbons than avoiding plastic bags.

Incidentally, a 120-page study by the Environment Agency in England [19] shows that conventional (HDPE) plastic bags have much less overall impact on the environment than the alternatives (biodegradable

---

[*] And no, we don't care about the difference between 10,000 and 15,000 here.

[†] If your car gets 10 or 40 mi/gal, feel free to multiply or divide our final numbers by two.

plastic, paper, or reusable cotton bags).* We just did a 1-page study that shows it is not worth doing a 120-page study.

Plastic bags are also a common form of litter. This is a big problem in some places. However, as a reader of this book, you carefully dispose of all your plastic bags and can therefore guiltlessly choose whichever type of bag is more convenient for you.

---

* That is not even including the need to wash those reusable bags to reduce bacterial contamination [20].

Paper doesn't grow on trees!

How far should we
be willing to walk
to recycle that one
hundred–page
report?

HINT: How much does a new ream of paper cost?
HINT: What fraction of the paper costs are raw material
costs?

ANSWER: Why are you throwing out that one hundred-page report? Paper doesn't grow on trees! Oh. Wait a minute. Paper *does* grow on trees. Maybe we should try to figure out the actual value of one hundred sheets of used office paper.

As usual, we can try different approaches. We can start from the cost of new paper; we can try to estimate the cost of the raw materials the used paper replaces; or we can look it up.

When paper is recycled, it is reduced to pulp and then processed to remove contaminants like ink and toner. This means we should estimate the cost of growing, harvesting, and processing trees. It will be much easier to start with the cost of new paper, which already incorporates all of that. A ream (five hundred sheets) of copier or printer paper costs about $3 (more than $1 and less than $10). Most of the cost of that paper will be processing the wood pulp into paper, shipping it, and selling it. The cost of the pulp (and hence its value when recycled) will be somewhere between 1% and 100% of the cost of the paper, so we will estimate 10%. This means that the pulp to make one hundred sheets of paper costs

$$C = \$3 \times 10\% \times \frac{100 \text{ sheets}}{500 \text{ sheets}} = \$0.06$$

Now let's compare to reality. Collected and sorted white office paper is worth about $100 per ton or $0.1 per kg. That means that we need to know the mass of the paper. We can estimate that by hefting a ream and estimating that it weighs a few pounds or a couple of kilograms. If we want to be more precise, we can estimate its volume and density. One ream of paper has a volume of about

$$V = 2 \text{ in} \times 8.5 \text{ in} \times 11 \text{ in} = 5 \text{ cm} \times 20 \text{ cm} \times 25 \text{ cm}$$
$$= 3 \times 10^3 \text{ cm}^3$$

and a density close to that of water, $\rho = 10^3$ kg/m^3 = 1 g/cm^3. The mass of a ream is then $m = 3 \times 10^3$ g = 3 kg. That one hundred–page report will have a mass of 0.5 kg. That implies its value is $0.05. Don't take any paper nickels.

If our time is worth $30 per hour (more than $10 and less than $100) or $20 per hour (the U.S. average GPD per capita of $40,000 divided by the 2,000 hours of work per year), $0.1 is worth 10 or 20 seconds. So it probably is worth recycling that thick report. But it is not worth worrying too much about 10 measly cents.

Note that our Sunday newspaper is probably worth about the same. It has two to three times the mass, but the paper is much lower quality.

The rain in Spain . . .

How much water
can we collect
from our rooftops?
How does this
compare to the
water we use
each year?

HINT: How much rain falls each year?

HINT: What is the area of the roof of a house?

ANSWER: In order to estimate the amount of water we can collect from our rooftops, we need to estimate the area of the roof and the amount of rain that falls each year. Because most Americans live in single-family homes, we will ignore apartments (for now). The average American single-family home has about 2,000 square feet of living space (more than 10^3 and less than 4×10^3) [17] and has two stories. This means that each home has about 10^3 square feet or 10^2 m^2 of roof. Now we just need the average annual rainfall. The majority of the U.S. population lives near a coast (East, West, or Gulf). I live on the East Coast, where our annual rainfall is about 1 m (40 inches). However, the annual rainfall in Los Angeles is less than half of that [21]. Let's use 1 m, as it is a nice round number. Feel free to adjust the results for your local rainfall.

This means that the annual volume of rain that falls on American homes (at least those east of Texas and the Dakotas or in the Pacific Northwest) will be about

$$V = Ah = (100\,\text{m}^2)(1\,\text{m}) = 10^2\,\text{m}^3.$$

At 10^3 L per m^3, this is a 10^5 liters or 2×10^4 gallons.

Now let's compare this to the water we use. We can estimate this several ways. We can consider the number of times we flush the toilet daily (at 3 gallons per flush), the length of time we spend showering (at 3 gallons per minute), the time spent watering the lawn (at 5 gallons or more per minute), the number of washing machine or dishwasher loads (at 10 to 20 gallons per load), etc. If we each flush the toilet three times per day, shower 10 minutes per day, water the lawn 5 minutes per day (more in the summer and less in the winter), and wash two loads each of clothes and dishes per week, that adds up to 70 gallons per person per day. Alternatively, we could look at our most recent water bill (and then remember how to convert from cubic feet to gallons).

Now we just need to convert from gallons per day to cubic meters per year. Here goes:

$$V = 70 \frac{\text{gal}}{\text{day}} \times \frac{400 \, \text{day}}{1 \, \text{yr}} \times \frac{4 \, \text{L}}{1 \, \text{gal}} \times \frac{1 \, \text{m}^3}{10^3 \, \text{L}}$$

$$= 10^2 \, \text{m}^3/\text{yr}.$$

This means that, even on the relatively wet East Coast, the roof of an average house can only collect enough water for one of its inhabitants. In drier Los Angeles the roof of an average house can only collect enough water for about a third of a person.[*]

---

[*] Or half a person, if the residents don't water the lawn.

How many oysters would it take to keep the
Chesapeake Bay clean and clear?

HINT: What is the volume of the Chesapeake Bay?

HINT: How much water does an oyster filter?

HINT: How fast (in m/s) does the oyster pass water over
itself? What is the water flow area?

HINT: How often should the Chesapeake Bay water be
filtered to stay clean and clear?

ANSWER: Oysters are filter feeders. They pass water over their gills and trap suspended particles (both edible and inedible) on the surface mucus, excreting the waste as feces. Thus, oysters efficiently remove suspended particles from water.

In order to estimate the number of oysters needed to keep the Chesapeake Bay water clean and clear, we need to estimate three quantities, the volume of the bay, the rate at which oysters process water, and frequency with which bay water should be cleaned. Here we go.

The Chesapeake Bay stretches from Norfolk and Virginia Beach in the south, past Washington and Baltimore, to almost the Mason-Dixon line (the Maryland/Pennsylvania border). If you're from the area, you might know that you can drive from one end to the other in 3 or 4 hours. Otherwise, we could estimate that the distance is about 10% of the Atlantic coastline (about 1,500 miles). In either case, we'll estimate the length is about 200 miles (300 km). The bay is relatively narrow, so the average width is about 30 km (more than 10 km and less than 100 km). The bay is also relatively shallow, with a depth between 10 ft and 1,000 ft. Taking the geometric mean gives a depth of 100 ft (30 m). This means that the volume of the bay is

$$V_{Bay} = (3 \times 10^5 \, m)(3 \times 10^4 \, m)(30 \, m)$$
$$= (3 \times 10^{11} \, m^3).$$

Hmmm. It looks like we will need a lot of oysters.

You probably never thought you would be called upon to estimate the flow of water through an oyster. However, as usual, we can place some reasonable bounds. The first step is to break the problem into two pieces: the speed of the water (in m/s or mph) and the cross-sectional area of the water flow from which food particles can be extracted. The water will flow relatively gently, so that the particles have time to get trapped in the mucus. This implies a flow that is slower than a brisk walk (5 mph or 2 m/s) but faster than a snail's pace (1 cm/s), giving an estimated flow of $v = 0.1$ m/s. The cross sectional area will be more than $(1 \text{ mm})^2 = 10^{-6} \text{ m}^2$ and less than $(1 \text{ cm})^2 = 10^{-4} \text{ m}^2$, giving an area $A = 10^{-5} \text{ m}^2$. This implies a volume flow rate of

$$F = vA = (0.1 \text{ m/s})(10^{-5} \text{ m}^2) = 10^{-6} \text{ m}^3/\text{s}$$
$$= 4 \times 10^{-3} \text{ m}^3/\text{hr},$$

or 4 liters per hour.

Now we need to estimate how often the bay water should be filtered to keep it clean and clear. Perhaps the bay should be cleaned as often as I vacuum my house (or as often as I should vacuum my house).[*] As usual, when we do not know the answer, we can at least bound the problem. Filtering once a day is probably too frequent, once a month is probably too infrequent, and once a year is definitely too infrequent. Taking the geometric mean of one and one hundred days gives a desired filtering period of ten days.

In ten days, one oyster will filter a volume of

$$V_{\text{oyster}}^{10 \text{ day}} = (10 \text{ days})(24 \text{ hr/day})(4 \times 10^{-3} \text{ m}^3/\text{hr})$$
$$= 1 \text{ m}^3.$$

This means that in order to filter the volume of the bay every ten days, we will need

$$N_{\text{oyster}} = \frac{V_{\text{Bay}}}{V_{\text{oyster}}^{10 \text{ day}}} = \frac{3 \times 10^{11} \text{ m}^3}{1 \text{ m}^3} = 3 \times 10^{11}.$$

That is rather a lot of oysters.

[*] This is a silly comparison, as the physical processes and rates of dirt accumulation in my house and the bay are quite different (I hope).

Of course, because the Chesapeake Bay is the largest estuary in the United States, we should calculate the oyster density rather than the absolute number of oysters. The area of the bay is $A_{\text{bay}} = (3 \times 10^5 \text{ m})(3 \times 10^4 \text{ m}) = 10^{10} \text{ m}^2$. Thus to clean the bay, we will need an average oyster density of

$$d_{\text{oyster}} = \frac{N_{\text{oyster}}}{A_{\text{bay}}} = \frac{3 \times 10^{11} \text{ oysters}}{10^{10} \text{ m}^2} = 30 \text{ oysters/m}^2$$

OK, that is still a lot of oysters.

You light up my life!

How does the cost of lighting using
incandescent light bulbs compare with
the alternatives, such as candles, compact
fluorescents, or LED "light bulbs"?

HINT: The total cost of a bulb includes the purchase
price, the installation cost, the operating costs, the change
in the heating and cooling costs, and the disposal cost.

HINT: The power rating of a bulb (e.g., 100 W) indicates
the power consumed, not the light emitted.

HINT: Incandescent light bulbs convert about 2% of the
power they consume to light.

HINT: Electricity costs about $0.1 per kilowatt-hour.

HINT: All of the power consumed by a light bulb ends up
as thermal energy.

HINT: An incandescent bulb lasts about 1,500 hours.

ANSWER: The total cost of a light bulb includes the purchase price, the installation cost, the operating cost, the change in the heating and cooling costs, and the disposal cost (not to mention the cost of all the stupid light bulb jokes such as "How many *Guesstimation* authors does it take to change a light bulb?").[*] Let's consider a standard 100-W incandescent light bulb and its alternatives over a period of 10,000 hours.

An incandescent light bulb heats its filament (usually tungsten) to a temperature of about 3,000 °C where it becomes white hot. This is the same process by which the sun emits light, although the surface of the sun is a bit warmer at 5,500 °C.[†] That is why incandescent light is similar to daylight. All of the power consumed by the light bulb is radiated; however, most of that power is radiated in wavelengths that we cannot see.

Incandescent light bulbs are cheap; they cost about $1 each. I've been buying light bulbs recently, so I know that their lifetime is 1,000 to 2,000 hours. If you haven't memorized light bulb lifetimes, then you can estimate it from the frequency with which you change light bulbs. Incandescent light bulbs need to be changed about yearly (less than monthly and more than decadely). If the bulb is only on in the evenings, then in one year it is on for

$$t = (4 \, \text{hrs/day}) \times (400 \, \text{days/yr}) = 1,600 \, \text{hrs/yr}.$$

Using $t = 10^3$ hours for ease of calculation, over its lifetime a 100-W bulb will consume energy

$$E_{100W} = Pt = (100 \, \text{W}) \times (10^3 \, \text{hr}) = 10^5 \, \text{W} \cdot \text{hr}$$

$$= 10^2 \, \text{kWh}.$$

---

[*] More than 0.1 and fewer than 10, so we'll estimate one.

[†] We should really use the Kelvin absolute temperature scale, where the temperature in Kelvin equals the Celsius temperature plus 273. However, this is needlessly precise for these high temperatures.

At approximately $0.1 per kWh, that $1 light bulb will consume $10 of electricity.

There are also installation and disposal costs. It takes about 5 minutes to install a single light bulb (removing the shade or cover, removing the old bulb and throwing it out, hunting in the closet for a new bulb, installing the new bulb, replacing the cover). If we value our time at our average hourly wage (more than $10 and less than $100), then it takes about $3 of our time ($30 × 5/60) to install a light bulb.*

Because the energy consumed by the incandescent light bulb dominates its cost, we should also consider the heating and cooling costs it incurs. In the winter, the power consumed by the light bulb helps heat the room. In the summer, the power consumed by the light bulb needs to be removed from the room by the air conditioner.

Let's consider winter first. If we use electric heat to warm our house, then the light bulb is just another way to convert electrical energy into heat and is basically free. If our heat source costs one-half of electric heating, then every dollar spent on electric light reduces our heating bill by $0.50.

Now let's consider summer. An air conditioner is a heat pump. You put energy into the air conditioner and it pumps energy out of your house. The energy efficiency ratio (the energy removed divided by the input energy) will be more than 1 and less than 10, so we will use the geometric mean of 3. This means that for every joule of energy used to run the pump, it

---

* Obviously, if we install more than one bulb at a time, it will take a lot less than 5 minutes per bulb. This is why smart companies wait until a certain percentage of the light bulbs in an area are burned out (say 20%) and then replace all of the bulbs in the area at the same time. The labor savings from replacing one hundred bulbs at a time vastly exceeds the value of the remaining bulb lifetime.

pumps 3 joules of heat out of the house.[*] Thus, in the summer the air conditioning needed to compensate for the light bulb's power will consume an extra 30%.

Phooey. Because the additional heating and cooling power consumption are both a fraction of the light bulb power consumption and because they offset each other, we will ignore all of it.

Including the initial cost, installation, and operating expenses, the total system cost of operating a 100-W light bulb for 10^3 hours (about a year) is $1 + $3 + $10 = $14, which we will round off to $10.

Now let's consider candles. One candle emits about as much light as a 4-W night light. This means that we will need at least twenty-five candles to produce as much light as a 100-W light bulb. A medium-size (2-cm diameter, 10-cm height) plain candle lasts between 1 and 10 hours. Thus, in order to replace a single 1,000-hour 100-W bulb, we will need

$$N_{candles} = \frac{100\,W}{4\,W} \times \frac{10^3\,hr}{3\,hr} = 10^4.$$

Even at $0.1 per candle, that will cost $1,000 (not to mention the added insurance expense after we burn down our house a few times). There's a reason why we no longer prefer candles for illumination.

Now let's look at other alternatives. Fluorescent bulbs emit light in two steps. Passing an electric current through mercury vapor causes the mercury to emit light at specific wavelengths, some in the ultraviolet and some in the visible range. The ultraviolet light is then absorbed and reemitted at visible

---

[*] The efficiency is measured by the EER, energy efficiency ratio of the thermal energy (heat) removed to the power consumed. A typical EER for a room air conditioner is about 10. Unfortunately, that ratio is in (BTU/hr)/watt. In order to convert to a real efficiency, we need to multiply by 0.3. This means that an air conditioner with an EER of 10 can remove 3 joules of thermal energy for each joule of electrical energy consumed.

wavelengths by specific fluorescent coatings on the inside of the bulb. The coatings are chosen to optimize some combination of the total light output (the efficiency), the light quality (whiter or bluer or warmer), and the cost. A compact fluorescent light bulb (CFL) is just a fluorescent bulb twisted into a compact shape so that it can replace standard incandescent bulbs. Because the light quality differs from daylight and depends on the coatings, some people dislike the light emitted by CFLs.

A CFL is several times more efficient than an incandescent bulb (the packaging will say things like "22 W provides the same light as a 75-W bulb") and significantly more expensive (about $5 to $10). It also claims to have a much longer lifetime of about 10^4 hours. This means a 100-W equivalent CFL will consume

$$E_{CFL} = \frac{100 \, W}{4} \times (10^4 \, hr) = 2.5 \times 10^5 \, W \cdot hr$$
$$= 300 \, kWh$$

during its lifetime. That will cost about $30 for 10^4 hours, or about four times less per hour than the incandescent bulb. The installation cost should be the same. However, because the CFL contains mercury, there are (or at least should be) extra disposal costs involved. These should be small compared to the total system cost of operating a CFL for 10^4 hours of $10 + $25 + $3 = $40. Thus compared to an incandescent bulb of the same brightness, it costs about three times as much to operate a CFL for ten times longer.

We should compare the lighting sources over a reasonable time period such as one year, not over their expected lifetimes. Assuming that a light is used for 1,600 hours per year, a 100-W incandescent bulb will cost $20, and the equivalent CFL will cost only $6. If we leave ten lights on for 4 hours a day, replacing incandescents with CFLs will reduce our lighting costs from \$200 to \$60 and save \$140 per year.

Comparing LED lights is more difficult because the technology is changing rapidly. Efficiencies are increasing, and prices are dropping.

However, by switching from candles to incandescents to CFLs (and that is not even considering whale oil), we have already reduced our lighting expenses (for the equivalent of ten 100-W incandescents) from 10^4 to \$200 to \$60 per year. Compared to the approximately 10^3 we pay for gasoline (500 gallons per year at \$2 per gallon) and the 10^3 we pay for heating and cooling our homes, the \$60 cost of CFL lighting is negligible.

At that point, cost is no longer a reason to switch lighting technology. If we switch to LED lighting, we will do it to improve light quality, to replace mercury-containing CFLs, and to increase bulb lifespan.

The Five Senses

Chapter 4

We view the world through our senses. Our eyes can see at noon and at midnight. Our ears can hear a whisper and a rock concert. Our senses have a remarkable range and sensitivity. Let's explore that in this chapter.

Don't stare at the Sun

4.1

What is the maximum amount of light that our eyes can tolerate, even briefly?

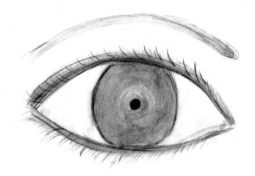

¿ ¿

HINT: What is the area of the pupil on a sunny day?

HINT: The power output density of the Sun is 10^3 W/m^2 at the Earth's orbit.

HINT: The answer should be in units of power or W.

ANSWER: We need to estimate the power output of the brightest light we can look at and the area of our pupil. Because we can look directly at the Sun (at least for very short periods of time),* the maximum power density we can tolerate is less than or equal to the solar constant, which is about 10^3 W/m^2.† At this level of light, the pupil is as constricted as possible so that it is 2–3 mm across, and its area is about 5 mm^2, or 5×10^{-6} m^2. This means that the power of the light entering the eye is

$$P = (10^3 \text{ W/m}^2)(5 \times 10^{-6} \text{ m}^2) = 5 \text{ mW}.$$

Note that this is about the same as the 5 mW maximum power of standard laser pointers. This is also why you should not shine laser pointers in people's eyes.

---

* Please don't try this at home.

† If we want to be a little more precise, then we should divide this by two to account for the power at the Earth's surface (at noon in the summer).

What is the minimum amount of light that our eyes can detect?

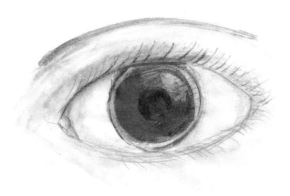

ċ ċ

HINT: Visible photons have an energy of about 1 eV.

HINT: Visible photons (unlike ultraviolet photons) do not have quite enough energy to cause most chemical reactions (like sunburns).

HINT: We can detect about 10 photons.

HINT: The answer should be in units of power or W.

ANSWER: In order to determine the minimum amount of light we can detect, we need to estimate the fewest number of photons[*] we can detect, the energy per photon, and the time interval. In order to see in the dark and not get eaten by predators,[†] the minimum number of photons should be as small as possible. In order to limit false alarms, the minimum number should be significantly more than one. Ten is a reasonable compromise between seeing things that are not there and not seeing things that are there. The minimum flash of light that we can perceive (after our eyes are fully dark-adapted) is about 10 photons.[‡]

Now we just need to estimate or calculate the energy of one visible photon. There are a few ways to do this. The simplest way is to remember that a chemical reaction is about 1.5 electron volts (eV), because a standard AA- or D-cell battery uses chemical reactions to achieve an electrical potential of 1.5 V. A visible photon cannot cause most chemical reactions, but an ultraviolet one can (think about sunburns). Thus, a visible photon will have an energy of about 1 eV.

To convert from the microscopic electron volts to the macroscopic joules, we use

$$1\,J = 6 \times 10^{18}\,eV.$$

This is a very useful number that we will use often. If there is still room on your shirt cuff, please record it there. Thus, we estimate that a visual photon will have

---

[*] Light is composed of particles called photons. We do not perceive individual photons because there are normally so many of them.

[†] Evolutionary pressures have changed somewhat in the past few years.

[‡] Back in the early days of nuclear physics, Ernest Rutherford's assistants detected subatomic particles by observing the tiny flash of light emitted when the particle struck a zinc-sulfide screen. That was when physicists were men of vision.
And yes. Back then they were almost all men.

energy

$$E_{\text{photon}} = \frac{1\,\text{eV}}{6 \times 10^{18}\,\text{eV/J}} = 2 \times 10^{-19}\,\text{J}.$$

The other method requires a little more specialized knowledge.* The wavelength of visible light is about $\lambda \approx 500$ nm (5×10^{-7} m). If we remember Planck's constant (and doesn't everyone?), $h = 7 \times 10^{-34}$ J-s, then we can use

$$E = h\nu = \frac{hc}{\lambda} = \frac{(7 \times 10^{-34}\,\text{J-s})(3 \times 10^8\,\text{m/s})}{5 \times 10^{-7}\,\text{m}}$$

$$= 4 \times 10^{-19}\,\text{J}.$$

Note that our first estimate is too low by a factor of two. Oops. The mistake is less than a factor of ten, so let's not lose too much sleep over it.

Now we need to estimate the resolving time of the human eye. We cannot resolve (that is, see) individual movie frames at 22 frames per second but we can resolve things that happen a few times per second. Thus, let's use a resolving time of about 0.1 s and assume that all ten photons must arrive in this time.

Now we can convert the photon energy to power. Ten photons in 0.1 s gives a minimum detectable power of

$$P = \frac{(10\,\text{photons})(4 \times 10^{-19}\,\text{J/photon})}{(0.1\,\text{s})}$$

$$= 4 \times 10^{-17}\,\text{W}.$$

Thus, the human eye (when properly dark- or light-adapted) can sense power levels from 5×10^{-3} W to 4×10^{-17} W. Wow! That is a range of 10^{14}, or 100 trillion. That is a truly amazing feat of engineering!

---

* That's a hint that you can safely ignore the rest of this paragraph.

Light a single candle

4·3

How far away can we see a candle burning on a clear, dark night? Assume we are far away from all other forms of illumination.

¿ ¿

HINT: How much light does a candle emit? Compare it to a 4-W night light.

HINT: How many photons per second is that?

HINT: How large is our pupil when fully dark-adapted?

ANSWER: To answer this, we need to estimate the light output of a candle, the sensitivity of the human eye when fully dark-adapted, and the light-collecting area of the eye. In the previous problem we estimated that the human eye can detect ten photons arriving in 0.1 second, corresponding to a power of

$$P = 4 \times 10^{-17} \text{ W}.$$

When completely dark-adapted, the human pupil is about 5 mm (almost 1/4 inch) in diameter, giving it an area of $A = 25 \text{ mm}^2 = 2 \times 10^{-5} \text{ m}^2$. This means that we can detect a power density of

$$p = \frac{4 \times 10^{-17} \text{ W}}{2 \times 10^{-5} \text{ m}^2} = 2 \times 10^{-12} \text{ W/m}^2.$$

If this power density covered the Earth, it would correspond to a total light power of $P = pA = (2 \times 10^{-12} \text{ W/m}^2)(4 \times 10^{14} \text{ m}^2) = 10^3 \text{ W}$. That is 1 square meter of sunlight spread out over the entire surface of the Earth!

Now let's estimate the light flux from a candle. A candle puts out about the same amount of visible light as a 4-W incandescent night light. The efficiency of an incandescent bulb is quite low, only a few percent. This means that the candle or 4-W night light emits only about 0.1 W of photons.[*]

Now we can estimate the maximum distance at which we can detect the candle. At a distance r from the candle, the power output of the candle is spread out over a sphere of radius r and area $A = 4\pi r^2$. Thus, the power density of the candle at a distance r

---

[*] If we remember the definition of a candela (which I just looked up), we would know that a common candle puts out $\frac{1}{683}$ W/steradian. With about 12 sr in a sphere, that amounts to a total light output of about 0.02 W, so our estimate is within a factor of five.

is $p = 0.1\,\text{W}/(4\pi r^2)$ so that

$$r = \sqrt{\frac{0.1\,\text{W}}{4\pi p}}.$$

Owing to use of a square root in this equation, errors of a factor of four in either the power output of a candle or the minimum power detectable by the human eye will cause an error of only a factor of two in the maximum distance. The maximum r will correspond to the minimum detectable p so that

$$\begin{aligned}
r_{max} &= \sqrt{\frac{0.1\,\text{W}}{4\pi\,(2 \times 10^{-12}\,\text{W/m}^2)}} = \sqrt{2 \times 10^9\,\text{m}^2} \\
&= 5 \times 10^4\,\text{m} \\
&= 50\,\text{km}.
\end{aligned}$$

This is rather a lot. At this distance, we will also need to include the effect of atmospheric attenuation. This will reduce the distance somewhat (depending on humidity and pollution). Note that at this distance, we will only be able to dimly perceive the candle by not looking at it directly.

This visibility of faint lights over very long distances is why cities subject to air raids in World War II were so careful about maintaining total blackouts.

Oh say can you see?

What is the maximum angular resolution of the human eye? What biological or physical characteristics limit this resolution?

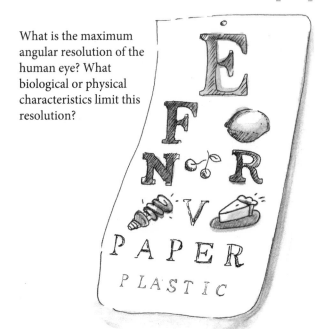

HINT: What is the smallest thing we can see at a distance of 1 m?

HINT: How large are the rod and cone cells in the retina?

HINT: How large is the retina?

ANSWER: The resolution of our eye refers to the smallest item we can see at a certain distance. The size of the item will depend on the distance. We might be able to see a mosquito on our arm but not the same mosquito 10 feet away on our friend's arm. In general, the shorter the distance, the smaller the object we can see.[*] Typically, if we can see a 1-mm object at a distance of 1 m, we can see a 10-mm object at a distance of 10 m and a 1-m object at a distance of 1 km. For objects that are much smaller than the distance to them, their angular size is simply $\theta = $ size/distance.[†]

We can estimate the angular resolution of the eye either by measurement, from biology, or from physics. Let's do all of them.

If we hold a meter stick at arm's length (about 1 m away), we can easily resolve the individual millimeters, each of which subtends an angle $\theta = 1$ mm / 1 m $= 10^{-3}$ rad, and we can also resolve the mm lines themselves, which are about ten times smaller. This means that the angular resolution of the eye is about 10^{-4} rad, or about 6×10^{-3} degrees.[‡]

The biological limit on the angular resolution of the eye is the number of pixels. The detectors in digital cameras are measured in pixels. That is the number of independent sensors in the focal plane of the camera. The more pixels, the better the resolution. The eye also has pixels, known as rod and cone cells, in the retina. The size of a typical cell is a few micrometers. We can estimate this size from looking through microscopes at cells. Because the wavelength of visible light is about 0.5μm,[§] anything we can see

---

[*] Except for those of us older folks with presbyopia, who cannot focus on close objects.

[†] The angle θ is measured in radians, where 1 radian ≈ 60 degrees.

[‡] That is about 20 arcseconds, for ancient Babylonians, astronomers, and others who use base 60.

[§] That is one of the useful numbers to write on your shirt cuff.

with a light microscope[*] must be at least that large. We can see structures about one-tenth the size of the cell. Therefore the cell is at least 10 wavelengths in diameter, or 5μm. (Note that the actual diameter of cone cells ranges from 0.5 to 4μm. They are about 50μm long.)

The diameter of the eyeball is about 1 inch, 3 cm, or 3×10^{-2} m. Thus, each cone cell subtends an angle on the back of the eyeball (the retina) of

$$\theta = \frac{2 \times 10^{-6}\,\text{m}}{3 \times 10^{-2}\,\text{m}} = 6 \times 10^{-5}\,\text{rad}$$

as seen from the pupil.

An alternative method to determine the angular extent of one cone cell is to measure the size of our visual field. As I type this in a classroom, I can see the entire 20-feet width of the rear wall (while staring fixedly at a single point), which is about 20 feet from me. (I am using English units, as my rulers are the 2-foot square ceiling tiles.) This is equivalent to standing at one point of an equilateral triangle and looking at the far side of the triangle, which is about 20 feet from me, so that my field of vision covers an angular range of about 60°. This 60° (or 1 rad) maps to about 2 cm of retina so that a 2×10^{-6}-m diameter cone cell covers

$$\theta = \frac{2 \times 10^{-6}\,\text{m}}{2 \times 10^{-2}\,\text{m}}(1\,\text{rad}) = 10^{-4}\,\text{rad}.$$

Thus, even if our lens and cornea could focus light perfectly, the size of our pixels limits our visual acuity to 10^{-4} rad.

The physics limit comes from the bending of light. Because light is a wave, it bends (diffracts) as it passes

---

[*] Or with a heavy one.

through any aperture.* Just as water waves spread out as they pass a barrier and sound waves spread out as they pass through a doorway, light waves spread out as they pass through an aperture. This means that, no matter how perfectly the lens and cornea focus the light entering the pupil to a single spot, that light will still spread out over some angle θ. The angular spread (diffraction limit) for any circular aperture is

$$\sin \theta \approx \lambda/d,$$

where θ is the amount that the light spreads out (in radians), λ is the wavelength of visible light, and d is the aperture of our optical instrument (e.g., our eye).

Trig functions are scary when we are trying to estimate. Fortunately, $\sin \theta \approx \theta$ when $\theta < 0.5$ rad (or $30°$), so we can replace $\sin \theta$ with θ.†

Visible light has a wavelength of between 0.4 and 0.7μm, so we will use $\lambda = 5 \times 10^{-7}$ m. The diameter of the pupil depends on the level of ambient light but is typically between 2 and 5 mm. Thus,

$$\theta = \frac{\lambda}{d} = \frac{5 \times 10^{-7} \text{ m}}{5 \times 10^{-3} \text{ m}} = 10^{-4} \text{ rad}.$$

All three methods agree. This is unsurprising, as we expect that evolution would optimize the number of pixels and the design of the eye to take advantage of the maximum physically possible resolution.

---

* Yes, I know that in previous questions we used the fact that light is a particle. Light is both a wave *and* a particle. In general, light travels as a wave and interacts as a particle.

† Mathematics students are taught that $\sin \theta = \theta$ when $\theta \ll 1$. However, $\sin(0.5) = 0.48$. Thus, even when θ is as large as 0.5, the difference between θ and $\sin \theta$ is only 4%. That is definitely close enough for this book.

While we're on the subject of optical resolution, what size telescope aperture is needed to see a possible Earth-like planet circling a nearby star?

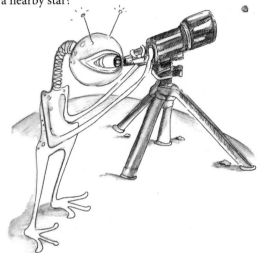

ċ ċ

HINT: The Earth-like planet will be about the same distance from its star as the Earth is from the Sun.

HINT: Nearby stars are a few light-years away.

HINT: Assume that we can collect enough light to see the planet, but that we need enough resolution to distinguish it from its star.

ANSWER: In order to see a planet circling its star, we need to have enough angular resolution that the two objects are well separated. There are two angular sizes involved. One is the angular separation between the planet and the star. This determines whether we can see both objects separately or whether they are smeared together. The other is the angular size of the planet. If our angular resolution is, say, one hundred times the size of the planet, then the already relatively dim light of the planet will be smeared out over a larger area and will appear even dimmer relative to the very bright star. Therefore, even if we have enough angular resolution to separate the planet and the star, we would still want to improve our resolution to better see the planet.

Let's start with the angular separation of the planet and the star by assuming that the planet is one Earth-orbit (1 astronomical unit [AU]) from its star and that the star is "nearby." This means that the planet is 93 million miles, or 1.5×10^{11} m, from the star and that the star is about 10 light-years from us. One light-year is the distance that light travels in one year. The speed of light is $c = 3 \times 10^8$ m/s, and the time is one year $= \pi \times 10^7$ s, so 1 l-y $= (3 \times 10^8$ m/s$)(\pi \times 10^7$ s$) = 10^{16}$ m. This means that the angular separation between the planet and its star is

$$\theta = \frac{\text{size}}{\text{distance}} = \frac{1.5 \times 10^{11} \text{ m}}{10 \times 10^{16} \text{ m}} = 10^{-6} \text{ rad.}$$

This is only one hundred times better than the resolution of the human eye. Given that the angular resolution $\theta \approx \lambda/d$, this means that we need a telescope with an aperture of

$$d = \frac{\lambda}{\theta} = \frac{5 \times 10^{-7} \text{ m}}{10^{-6} \text{ rad}} = 0.5 \text{ m,}$$

which is one hundred times larger than the human eye. That is not much. The 5 m telescope on Mt. Palomar was built more than sixty years ago but has never directly seen planets. This estimate is clearly much too small.

There are two problems with this estimate. The first is that the diffraction formula we used, $\theta \approx \lambda/d$, applies only to separating two objects of similar brightness. If one object is much brighter than the other, then we need much more separation to distinguish them. The second problem is that because the planet is so dim, we want to collect a lot of light and also minimize the size of its image to concentrate that light in a small region.

Ideally, the telescope's resolution would equal the angular size of the planet. Because the diameter of Earth is $d = 6 \times 10^6$ m, the ideal angular resolution is

$$\theta = \frac{6 \times 10^6 \text{ m}}{10 \times 10^{16} \text{ m}} = 6 \times 10^{-11} \text{ rad.}$$

This is about 1 million times better than the resolution of the human eye. This implies a telescope aperture of

$$d = \frac{\lambda}{\theta} = \frac{5 \times 10^{-7} \text{ m}}{6 \times 10^{-11} \text{ m}} = 10^4 \text{ m.}$$

Hmmm. A 10-km aperture telescope. This could be difficult. The largest telescope in use today is about 10 m. Fortunately, this is a goal rather than a requirement.

This indicates the difficulties of directly imaging Earth-like planets circling other stars at Earth-like distances. As of the time of this writing, while scientists have detected many exo-planets, they seen only have a few Jupiter-like planets directly.

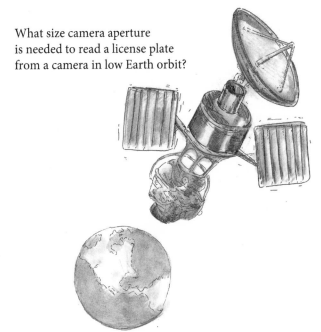

They're watching us!

What size camera aperture
is needed to read a license plate
from a camera in low Earth orbit?

ċ ċ

HINT: Assume that the camera lens focuses perfectly so
that the resolution is only limited by the diffraction of light
as it passes through the aperture.

HINT: How far away is the camera?

HINT: How thick are the numbers and letters on a license
plate?

ANSWER: In order to determine the camera aperture, we need to estimate the angular resolution required to read the license plate. This means that we need to estimate the distance from the camera to the license plate and the spatial resolution required to read the license plate.

Let's start by estimating the height of a satellite in low Earth orbit (LEO). There are several ways to do this. We might remember that LEO is about 300 km (200 mi) high. We might remember that outer space is defined as starting at 100 km above the surface. If we don't remember either of those, we can consider the atmosphere.

In order to stay in orbit, the satellite must be above almost the entire atmosphere so that there is almost no air to slow it down. Let's estimate the height of the atmosphere. The density of air on the top of Mt. Everest is about one-half the density of air at ground level (this is why most climbers on Mt. Everest use oxygen bottles). Mt. Everest is 10 km tall. Assuming that the density of air decreases by a factor of two for every 10 km, this means that it decreases by a factor of $2^{10} \approx 10^3$ for every 100 km. Thus it decreases by a factor of 10^6 at 200 km and by a factor of 10^9 at 300 km.

Let's be safe and place our satellite 300 km above the Earth's surface.

Now let's estimate the resolution needed to read a license plate. The license plate letters and numbers are about 1 cm thick, so 1-cm resolution should be good enough. Alternatively, the display size needed to create readable characters is approximately five pixels by eight pixels. Because license plate characters are about 2 inches (5 cm) by 4 inches (10 cm), 1-cm resolution would imply a 5- by 10-cm display and should be quite readable.

This means that we need an angular resolution of

$$\theta = \frac{\text{size}}{\text{distance}} = \frac{10^{-2}\,\text{m}}{3 \times 10^5\,\text{m}} = 3 \times 10^{-8}\,\text{rad},$$

or about 10^4 times better than the human eye. In order to achieve that resolution, we need a camera aperture of

$$d = \frac{\lambda}{\theta} = \frac{5 \times 10^{-7}\,\text{m}}{3 \times 10^{-8}\,\text{rad}} = 20\,\text{m}.$$

The Hubble Space Telescope has a 2.4 m aperture. This camera aperture is on the boundary of the possible. Now we would need to do a lot more work to determine the precise aperture needed. Among other things, we would have to figure out the exact spatial resolution needed to read license plates; the exact height of the satellite; and the exact relationship between aperture, resolution, and wavelength. All that is way beyond the scope of this book.

This is one of those few questions where the answer is in the "just right" Goldilocks category.

Oh well.

Beam the energy down, Scotty!

How large (in area) a receiving antenna would
we need to receive the power beamed down
using microwaves from a 1-GW (10^9 W)
solar-power satellite in geostationary orbit?

HINT: This is just another question dealing with
electromagnetic waves and apertures.

HINT: Geostationary satellites orbit about 4×10^4 km
from Earth's center.

HINT: How much area would the satellite need to collect
all that solar energy?

HINT: The solar power density at Earth orbit is about
10^3 W/m^2.

HINT: Microwaves have a wavelength of about 1 cm.

HINT: How much would the beam from the satellite
spread out as it traveled to Earth?

ANSWER: This is just another question dealing with electromagnetic waves and apertures. The microwaves will be beamed down from a radio-dish antenna mounted on the satellite. Because the beam will spread out as it travels, we will need a large receiving antenna on the Earth. The radio-dish antenna aperture will determine how much the beam spreads, using the now-familiar formula, $\theta = \lambda/d$, where λ is the microwave wavelength and d is the diameter of the radio-dish. The diameter of the beam on its target[*] is then the angular spread times the distance traveled. This means that we need to estimate the diameter of the solar-power satellite's antenna, the wavelength of microwaves, and the distance from the satellite to the Earth.[†]

This problem is a little more complicated than the others, but we should be able to handle it now. Here we go.

Let's start with the size of the solar-power satellite. We want to collect 1 GW of power. Given that the solar power density at Earth orbit is 10^3 W/m^2, at 100% efficiency we will need an area of

$$A = \frac{10^9 \text{ W}}{10^3 \text{ W/m}^2} = 10^6 \text{ m}^2 = 1 \text{ km}^2.$$

However, the efficiency will be a lot less than 100% and a lot more than 1%. Let's take the geometric mean and estimate 10%. This means that our solar-power satellite will need ten times more area, or

$$A = 10^7 \text{ m}^2 = 10 \text{ km}^2.$$

That is a lot of area and significantly more than we can achieve today.

---

[*] In this case, it is the Earth; in the case of the eye, it is the retina.

[†] Note that the beam from a larger transmitting antenna will spread less. This means that if we increase the size of the transmitting antenna, we can decrease the size of the receiving antenna.

In order to estimate the wavelength of microwaves, we can use several different methods. The most fun method is to spread popcorn kernels evenly on several glass or plastic trays. Place the trays at different heights in a microwave oven (without using the turntable). Then run the oven until about a quarter of the kernels have popped. They will pop unevenly, with popped spots and unpopped spots. The different spots will form an interference pattern. The average distance between popped spots is about the wavelength of the microwaves.

Let's try to bound the wavelength without measuring it. Microwave wavelengths are significantly longer than visible or infrared light (because if the wavelengths were shorter than visible, they could cause burns, as ultraviolet light does). Thus the wavelength will be much longer than 10^{-6} m. Microwave wavelengths will also be significantly smaller than the size of the microwave oven (0.3 m). Taking the geometric mean of these bounds, the microwave wavelength is $\lambda = \sqrt{(0.3\,\mathrm{m} \times 10^{-6}\,\mathrm{m})} = 5 \times 10^{-4}$ m, or about half a millimeter. However, these bounds are very far apart and do not give us much confidence in the resulting estimate.

Let's go back to that interference pattern. When there are two sets of waves with the same wavelength in the same space, they create an interference pattern. We can see this with water waves when we drop two rocks into the same pond. When a wave bounces off a surface and the incident and reflected waves pass through the same space, they also interfere. This interference creates regions of high intensity and low intensity. The distance between these regions is typically the wavelength of the wave. If the microwave wavelength is only 0.5 mm, then the distance between "hot" and "cold" regions in the microwave caused by the interference pattern would only be about 0.5 mm.

This is so small that microwave oven turntables would be unnecessary. In order for turntables to be useful, the distance between hot and cold regions must be at least a few centimeters. This lets us improve our wavelength bounds. The upper limit from the size of the microwave oven is still 30 cm, but the lower limit is now 1 cm. So we can take the geometric mean with more confidence to give an estimated wavelength of 5 cm.[*]

Now we need the distance from the satellite to the Earth. The satellite is in geostationary orbit. This means that it orbits the Earth once every day so that it stays directly above the receiving antenna. We can estimate this distance in a few different ways. We might remember that it is about 25,000 miles up. We might be able to use the equations of orbital motion to solve for the orbit. We might remember Kepler's relationship that the orbital period squared is proportional to the radius cubed ($T^2 \propto r^3$).

However, we can also estimate it. There are two orbits that should be somewhat familiar. A satellite in low Earth orbit (LEO) has a period of about 90 minutes (1.5 hours or $\frac{1}{16}$ day) and orbits about 6×10^6 m from the center of Earth. The Moon has a period of about 30 days and orbits about 4×10^8 m from the center of the Earth. Because the geometric mean of those two periods ($\frac{1}{16}$ day and 30 days, or 1.5 hours and 700 hours) is close to one day, the geometric mean of the distances should be close to that of a geostationary orbit. Thus, we estimate that the orbital radius is

$$r_{\text{geo}} \approx \sqrt{(6 \times 10^6 \text{ m})(4 \times 10^8 \text{ m})} = 5 \times 10^7 \text{ m},$$

or 5×10^4 km.

---

[*] We could also use 6 cm, but 5 makes the arithmetic easier.

Now we have all the ingredients for the answer. Let's assume that the transmitting antenna is as large as our 10-km^2 solar-power satellite, with a diameter of 3 km. The beam's angular spread is $\theta = \lambda/d$, and the spatial spread of the beam is $s = r\theta$, so we have:

$$s = \frac{r\lambda}{d} = \frac{(5 \times 10^7 \text{ m})(5 \times 10^{-2} \text{ m})}{3 \times 10^3 \text{ m}}$$
$$= 8 \times 10^2 \text{ m}$$
$$= 1 \text{ km}.$$

Thus, our receiving antenna would need to have an area of about 1 km^2, or about ten times smaller than the transmitting antenna. The microwave power density at the antenna will be ten times larger than the solar power density, or about 10 kW/m^2.

This could be a problem. We might want to spread the beam out more so that we don't fry birds and airplanes. Fortunately, it is much easier to spread the beam out than to focus it in.

Oh say can you hear?

What is the maximum angular resolution of the human ear? How well can we spatially locate things by sound alone (horizontally, not vertically)? What biological or physical characteristics limit this resolution?

ĉ ĉ

HINT: Imagine being out in an open field with our eyes closed. How closely could we point to a barking dog?

HINT: Assume a reasonable pitch (not too high and not too low).

ANSWER: As usual, we can estimate our aural angular resolution in different ways, either from measurement or from physics. Imagine standing in an open field with our eyes closed. A dog barks. How closely can we point to the dog just using sound? We can certainly localize the sound much better left to right than front to back, so let's just consider sounds coming from in front of us. Rather than trying to estimate angles in degrees or radians, let's rephrase the question slightly.

If there are a lot of dogs spread out evenly in a half-circle in front of us, how many could we distinguish among? We could certainly distinguish between two dogs and certainly could not distinguish among one hundred. We could probably distinguish among ten and probably not distinguish among forty. This means that we can distinguish about twenty items within a $180°$ range, giving an angular resolution of $10°$, or about 0.1 radians. This is 10^3 times worse than our optical angular resolution. Note that this is the horizontal resolution; we are much much worse at localizing items vertically.

Now let's consider the physical limitations. Our eyes have lots and lots of spatial sensors (rod and cone cells) but only three different frequency (color) sensors (red, green, and blue cone cells). Our ears have only two spatial sensors (two ears) but lots and lots of different frequency (pitch) sensors.

In order to localize a sound, our two ears exploit differences in the sound's arrival time and loudness. Consider the arrival time first. Sound travels in air at 300 m/s. Our ears are separated by about 8 inches, or 20 cm (0.2 m). If a sound originates directly in front of us, it will arrive at both ears simultaneously; if it originates off to one side, it will arrive at the farther ear

$$t = \frac{0.2\,\text{m}}{(300\,\text{m/s})} = 6 \times 10^{-4}\,\text{s} = 0.6\,\text{ms}$$

after it arrives at the nearer ear.* Distinguishing twenty angles ranging from the left (when the left ear's signal precedes the right ear's by 0.6 ms) to the right (when the right ear's signal is earlier by 0.6 ms) implies that the brain can resolve time differences of 0.06 ms, or 60 microseconds. Wow.

For higher frequency sound, the brain relies on sensing the loudness difference at the two ears. This will be a complicated function of head shape, ear shape, and frequency. This is why it is much harder to localize high pitched sounds.

---

* Note that what we are really sensing is not the difference in arrival times, as that only applies to the start of the sound, but to the phase difference. The phase difference is the difference in the arrival time of a single wave. This only works when the wavelength, λ, is larger than the ear separation. We can relate the speed of sound, c; the wavelength, λ; and the frequency, ν, by remembering the bad joke: "What's nu? c over lambda!" Thus, this only works when $\nu \leq (300 \, \text{m/s})/(0.2 \, \text{m}) = 1,500$ Hz. We don't need to worry about these details; this footnote is really here to deflect snide comments from experts.

Heavy loads

What is the dynamic range of our sense of touch? In other words, what is the ratio of the maximum tolerable weight to the minimum perceptible weight?

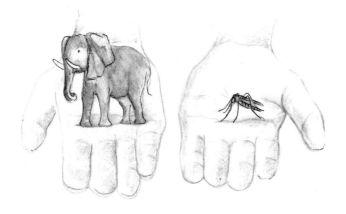

HINT: What is the lightest object we can detect?
HINT: What is the heaviest weight we can withstand?

ANSWER: Let's start by estimating the maximum weight we can withstand. A typical human mass of 100 kg corresponds to a weight of 10^3 newtons (N). We can tolerate the weight of a 100 kg person sitting on our lap. Ten 100-kg people on our lap would be too much. Alternatively, the maximum allowed acceleration on a roller coaster is $3.5g$ (as described in an exhibit on speed seen recently in the Pacific Science Center in Seattle).[*] Fighter pilots are subjected to accelerations up to $10g$. A brief acceleration of $100g$ (as in a car crash) will rupture the aorta.[†] Let's use a maximum safe acceleration of $10g$, which gives a maximum force of

$$F = 10mg = 10 \times (10 \text{ N/kg}) \times (100 \text{ kg}) = 10^4 \text{ N},$$

or 1 ton.

Now let's estimate the minimum detectable force. We can sometimes feel a mosquito land, but can rarely feel it once it has landed (until it "bites"). We can definitely feel the weight of a fly. Now we just need to estimate the weight of insects. A squooshed (that's a technical term) mosquito would occupy a volume of between 1 mm and 10 mm^3. At a density of 1, that gives a mass of between 1 mg and 10 mg. We'll take the geometric mean of 3 mg. A housefly is about one hundred times larger than a mosquito and has a mass of 100 mg. Because we can rarely detect the weight of 1 mg and can always detect the weight of 100 mg, we can take the geometric mean and estimate that the minimum detectable force is the weight of 10 mg.

---

[*] The variable g refers to the acceleration of gravity, so $3.5g$ means that the roller coaster exerts a force on our body that is 3.5 times stronger than that of gravity or 3.5 times our own weight.

[†] Force tolerance is strongly dependent on direction. The greatest tolerance is for "eyeballs in" forces. For incredibly more information, see NASA STD-3000, Man-Systems Integration Standards [22], especially section 5-3.

Thus, the smallest force we can detect is about

$$F_{\min} = mg = 10^{-2}\,\text{kg} \times 10\,\text{N/kg} = 0.1\,\text{N}.$$

This means that the dynamic range of our sense of touch is $10^4\,\text{N}/10^{-1}\,\text{N} = 10^5$. This is not bad, but it pales in comparison with the 10^{14} dynamic range of our eyes.

Energy and Work

Chapter 5

Now let's look at human energy sources.

How much energy can a human produce?

Is it really worth it?

The words "energy" or "energetic" are used loosely to describe many things, from performances to batteries and from toddlers to automobiles. In science, energy has a more restricted meaning, but it includes energy of position (potential energy), energy of motion (kinetic energy), thermal energy, chemical energy, gravitational energy, etc. In a previous chapter, we looked at energy and recycling. In this chapter, we will look at energy and motion.

There are several forms of potential energy, including gravitational and spring. Gravitational potential energy near the surface of the Earth is equal to an object's mass (in kg) times its height (in meters) times the gravitational constant $g = 10$ N/kg or

$$U = mgh.$$

The energy of motion, or kinetic energy, depends on the mass of the object and its speed:

$$K = \frac{1}{2}mv^2.$$

Similarly, the term "work" is used loosely to describe many activities. We will use its scientific definition: a form of energy transfer, specifically equal to the force applied times the distance over which it is applied,

$$W = Fd.$$

If the force and distance are in the same direction (as when we help push our friend's broken-down car off the road), then the work done is positive. If the force and distance are in opposite directions (as when we are pulling back on our dog's leash as he tries to catch that darned squirrel), then the work done is negative. If the force and distance are perpendicular to each other (as when the Earth pulls on the Moon in its orbit or when we swing a ball on a string around our head), then the work done is zero.

Force is measured in newtons ($1\,N = 0.2\,lb = 1$ apple); distance is measured in meters ($1\,m = 3\,ft$); and work and energy are measured in joules ($1\,J = 1\,N \times 1\,m$).

Similarly, "power" will mean the energy used per second, or the work done per second, and will be expressed in joules per second or watts ($1\,W = 1\,J/s$).

Energy can also be measured in kilowatt-hours. Because $1\,kW = 10^3\,W$ and 1 hour $= 3.6 \times 10^3\,s$, $1\,kWh = 4 \times 10^6\,J$. 1 kilowatt-hour of electricity typically costs around \$0.10 in the United States.

How much power
do we use as we run
up six flights
of stairs?

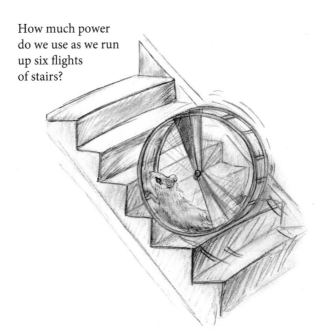

HINT: By how much does our potential energy change
from the bottom to the top?

HINT: What is our mass? Remember that 2 pounds is
approximately the weight of 1 kilogram.

HINT: How tall is a six-story building?

HINT: How much time would it take us to run up the
stairs?

ANSWER: In order to estimate the power used as we run up six flights of stairs, we need to estimate how much our potential energy will change and how much time it will take. This means estimating our mass, the height, and the time. The more mass, the more power. The higher the building, the more power. The more time, the less power.

My weight is about 180 pounds, which corresponds, in round numbers,[*] to a mass of 100 kg. Feel free to use your own weight. If you don't like running up stairs, use somebody else's weight.

Six flights of stairs is six floors. At about 10 to 15 feet per floor, that is 72 feet, or 24 m, which we will round off to 20 m.

It will take more than 10 s and less than 10^3 s (20 minutes) to run up six flights of stairs. Taking the geometric mean, we get 10^2 s, or 2 minutes. This seems reasonable. Feel free to go faster or slower if you prefer.

That means that we are exerting a power

$$P = \frac{mgh}{t} = \frac{(10^2 \, \text{kg})(10 \, \text{N/kg})(20 \, \text{m})}{10^2 \, \text{s}}$$
$$= 200 \, \text{W}.$$

Is this a lot? We need to compare it to a standard. A diet of 2,500 food calories per day corresponds to a metabolic rate of 100 W. Thus, 200 W is a significant extra exertion (no kidding!), but not impossible.

---

[*] I made that bad joke in the previous book. I like to think that I learned something from the experience.

How much energy would the United States generate per year if we connected all of our stair-steppers, rowing machines, treadmills, etc. to electrical generators? How much money would one person save by generating his or her own electricity that way?

HINT: Were you out of breath after running up the six stories in the previous question? Could you have kept going at that pace?

HINT: How many light bulbs did you manage to light on that science museum exercycle?

HINT: A bright incandescent bulb uses 100 watts.

HINT: 2,500 food calories per day equals 100 watts.

HINT: How much time each week do people spend working out?

ANSWER: To answer this we need to estimate the power output per person, the time spent working out, and the number of people working out. There are a few ways to estimate the power output. We can visit a science museum that has a stationary bicycle connected to a generator and electric light bulbs. The brightest possible bulb is almost always 100 W. (And it is almost impossible to keep that bulb lit for more than a few seconds.) We can start with our basic metabolic rate of 100 W (2,500 kcal per day) and assume that vigorous exercise would double that. We might know that 1 horsepower equals 760 W and assume that humans provide much less power than horses. Or we can use the answer from a previous problem. All of these methods will give answers that are definitely greater than 10 W, and less than 10^3 W, so we will use 100 W.

The average exerciser probably works out about 3 hours per week (more than 1 hour and fewer than 10 hours). The proportion of the population that works out is about 10% (more than 1% and less than 100%), or 3×10^7 Americans.

This means that the electrical energy that we could generate per year would be

$$
\begin{aligned}
E &= (100\,\text{W}) \times \left(3\frac{\text{hr}}{\text{wk}}\right) \times \left(50\frac{\text{wk}}{\text{yr}}\right) \times (3 \times 10^7) \\
&= 5 \times 10^{11}\,\text{W}\cdot\text{hr/yr} = 5 \times 10^8\,\text{kWh/yr} \\
&= 2 \times 10^{15}\,\text{J/yr} \, .
\end{aligned}
$$

This looks like a large number. It is a large number. It is equivalent to 100 kilotons of TNT. But let's look at it in a little more detail.

Let's assume that we are one of the hard-working people who works out 3 hours per week. In one week, we will generate 300 W·hr of electricity or 0.3 kWh. At $0.10 per kWh, we will save approximately two cents.[*] This will hardly pay for the generator needed to convert our hard work into electrical energy.

Despite this harsh economic reality, there is now a hotel that encourages its guests to generate energy on its exercise bikes and even rewards them with a free meal for producing 10 W·hrs of electrical energy. Wow! That's worth at least a tenth of penny! [23].

Thanks to Chuck Adler of St. Mary's College in Maryland for the question.

[*] OK. Three cents. But then we can't make bad jokes about putting our two cents in.

How much hydroelectric power could be generated from Niagara Falls?

HINT: By how much does the potential energy of the water change from the top to the bottom?

HINT: How tall is Niagara Falls?

HINT: How much water flows over the falls each second?

HINT: The density of water is 10^3 kg/m³.

HINT: How wide and deep is the Niagara River?

HINT: How fast does it flow?

ANSWER: Hydroelectric plants generate electrical power by converting gravitational potential energy to electrical energy. Because gravitational potential energy ($PE = mgh$) depends on mass and height, we need to estimate the rate of water flow over Niagara Falls and the height of the falls. As usual, we will estimate lower and upper bounds for each quantity and take the geometric mean for our estimate.

We will break the problem down to estimating the water width, depth, flow rate, and falls' height. Based on memories of paintings and photos of the falls, the river width at the top of the falls is more than 10 m and less than 10 km, so we will use the geometric mean of 300 m. The river depth is more than 1 m and less than 100 m, giving an estimate of 10 m. The flow rate is more than 1 m/s (2 mph) and less than 10 m/s, so we will use 3 m/s. The height of the falls is more than 10 m and less than 1 km, so we will estimate 100 m. This means that the mass of water flowing over the falls each second is

$$m/t = (300\,\text{m})(10\,\text{m})(3\,\text{m/s})(10^3\,\text{kg/m}^3) = 10^7\,\text{kg/s},$$

or 10^4 tons per second. Wow. That is a lot of water over the dam (or the falls, as the case may be).

The power that can be generated from this potential energy is

$$
\begin{aligned}
P &= \frac{m}{t}gh \\
&= (10^7\,\text{kg/s}) \times (10\,\text{m/s}^2)(100\,\text{m}) \\
&= 10^{10}\,\text{W} = 10\,\text{GW}.
\end{aligned}
$$

This seems high. One large coal- or nuclear-powered electrical plant generates "only" about 1 GW, one-tenth of our estimate for Niagara Falls.

According to the New York Power Authority, 1.5×10^6 gal/s (6×10^6 L/s or 6×10^3 m³/s) flows from Lake Erie into the Niagara River. The height of the falls is 52 m. Our (OK, it's really "my") flow rate and height estimates are each too high by a factor of two. In addition, at least half of the water cannot be used to generate power but must flow over the falls during daylight in tourist season.

Looking at these figures in detail, we see that we underestimated the width of the falls by a factor of three but greatly overestimated the water depth at the edge of the falls. Fortunately, these errors partially offset each other.

However, we are not actually that far off. Our estimate is only four times higher than the 2.4 GW that the NYPA power plant generates. This is well within our goal of a factor of ten.

That is the equivalent of more than two large coal- or nuclear-powered electrical plants. That's a lot of power!

Thanks to Albert J. Read of the SUNY College at Oneonta and the Science Discovery Center of Oneonta for suggesting the question.

A hard nut to crack

How much energy (work) does it take
to crack a nut?

HINT: How much distance do we apply it over?
HINT: How much force do we have to apply?

ANSWER: Let's consider an English walnut. Almonds are too easy, and black walnuts are much too difficult. Let's also use a nutcracker (unless we want to show off by cracking the nut with our bare hands). We will need to estimate the force applied and the distance traveled. The distance is easier. It will be a small fraction of an inch, or about 0.5 cm (because 1 cm = 0.4 in).

The force applied equals our grip strength. When we lift a heavy object, we close our hands first and then lift with our arm muscles. However, try to imagine the heaviest object we can lift with just our finger muscles. It is more than 1 lb and less than 100 lbs, so we'll take the geometric mean and estimate 10 lbs, or 50 N.

Now we can calculate the work done:

$$W = Fd = (50\,\text{N})(0.5 \times 10^{-2}\,\text{m}) = 0.3\,\text{J}.$$

That is not much energy at all. It is difficult to crack a nut because it takes a lot of force, not because it takes a lot of energy. In fact, that is why we (or at least, I) use a nutcracker. The nutcracker is a lever that allows us to use less force by increasing the distance over which we apply it.

Mousetrap cars

Could we power a car by using mousetraps for
energy storage (instead of gasoline or batteries)?
How much energy could we store in mousetraps?
How far could a mousetrap-powered car travel?

HINT: How much work does it take to "load" a
mousetrap?

HINT: What is the size of a mousetrap?

HINT: How many loaded mousetraps could we fit into a
car?

ANSWER: In order to figure out the practicality of using mousetraps to power our cars, we need to estimate how much mousetrap energy we can store in a car. To do this, we need to estimate a typical mousetrap's volume (or mass) and stored energy.

A traditional mousetrap has a spring-loaded wire bail that snaps shut when triggered. It is about 2 inches by 4 inches in area. The wire bail is about 2 inches long and thus requires 2 inches of vertical clearance. Because 2 inches = 5 cm, this means that the volume is

$$V = 5\,\text{cm} \times 10\,\text{cm} \times 5\,\text{cm} = 250\,\text{cm}^3.$$

Each trap will have a mass that is only an ounce or two, or a few tens of grams.

We'll need a lot of traps to power our car, so let's calculate the number of traps we can fit in a cube 1 m (3 ft) on each side. The volume of that cube is $V = 1\,\text{m}^3$. The volume of one mousetrap is $250\,\text{cm}^3$. Because 1 m = 10^2 cm, $(1\,\text{m})^3 = (10^2\,\text{cm})^3 = 10^6\,\text{cm}^3$. This means that we can fit

$$N = \frac{10^6\,\text{cm}^3}{250\,\text{cm}^3} = 4 \times 10^3$$

mousetraps in a cubic meter. (Another way to look at this is that we can fit twenty rows by ten columns of mousetraps into a square meter, and then we can stack them twenty high. This also gives us four thousand mousetraps per cubic meter.)

We can estimate the stored energy by estimating the force needed to set the trap. That force is more than 1 lb and definitely less than 100 lb, so let's use 10 lb (50 N). As this force is exerted over a distance of 10 cm, the work done[*] is

$$W = Fd = (50\,\text{N}) \times (0.1\,\text{m}) = 5\,\text{J}.$$

[*] Physics pedants will complain that I should use an average of half the force, but that is needlessly precise for this book.

Thus, the energy stored per volume is

$$E / V = (5 \text{ J}) \times (4 \times 10^3 \text{ mousetraps/m}^3),$$
$$= 2 \times 10^4 \text{ J/m}^3,$$

and the energy stored per mass is

$$E / m = \frac{5 \text{ J}}{30 \text{ g}} = 200 \text{ J/kg}.$$

Is this a lot or a little?

In order to have as much energy as possible, let's use a cubic meter of mousetraps or about ten times the volume of a typical gas tank. Those 4×10^3 mouse traps will have a mass of only 100 kg (that's the mass of 100 liters or 25 gallons of gasoline) and can store 2×10^4 J.

Gasoline has an energy density of 3×10^7 J/L, so our mousetraps will give us the energy of about 1 mL (10^{-3} L or 1 cubic centimeter) of gasoline. However, we do have one small advantage. The typical efficiency for converting the chemical energy in gasoline to mechanical energy is only about one-third. As the energy of mousetraps is already mechanical, one cubic meter of mousetraps will have the same useful energy as a whopping 3 mL (or half a teaspoon) of gasoline.

Now let's see how far we can go on 3 mL of gasoline. Thirty miles per gallon is about 50 km/gal, or 12 km/L, which, dividing both the numerator and the denominator by 10^3, is 12 m/mL. Thus our mousetraps will allow the car to travel approximately 40 m (120 ft) before we need to reset all of the traps.

Note also that if it takes 1 second to reset each trap, that is an energy transfer rate of $P = 5$ J/s = 5 W. This is a little less than the 10 MW we achieve at the gas station. About a million times less.[*]

[*] See *Guesstimation* for a discussion of gasoline energy density and energy transfer rates.

Looking at it another way, at 1 second per trap we can reset all the traps and "refuel" the car in a mere 4×10^3 s, or about 1 hour. This means that we will travel about 40 m, or 100 feet, per hour.

Oh well, back to the drawing board.

Thanks to Alex Godunov of Old Dominion University in Virginia for the question.

Push hard

How much work does it take to push
a car off the road?

Ⴤ Ⴤ

HINT: How much force does it take to get the car moving?

HINT: How much force does it take once the car is
moving?

HINT: How far do we push it?

HINT: Why isn't our friend pushing harder? It's his car.

ANSWER: We need to estimate the force applied and the distance traveled. We have to apply a large force to start the car moving. However, once it is moving, it is easy to keep the car moving.[*]

In order to start the car moving, we need to push as hard as we can. That is about 100 lbs of force (or about 500 N). Fortunately, we only need to push that hard for about 10 ft (or about 3 m). After that, we only need to apply a fraction of that force, say about 100 N (20 lbs). We can probably get the car to the side of the road in about only 10 m (30 ft). This means that the total work done is

$$W = Fd = (500\,\text{N})(3\,\text{m}) + (100\,\text{N})(10\,\text{m})$$

$$= 3 \times 10^3\,\text{J}.$$

Let's see if this is reasonable. We estimated that it takes about 2×10^3 J to get the car rolling. At that point, it is moving at walking speed or about 1–2 m/s (2–4 mph). The energy of motion is $K = \left(\frac{1}{2}\right) mv^2$. A small car with a mass $m = 10^3$ kg (1 ton) and a speed $v = 1$ or 2 m/s has a kinetic energy between 500 and 2,000 J. Close enough.

Note that this is less than the energy stored in all those mousetraps.

---

[*] Maybe I could say that better. How about: "An object at rest tends to stay at rest, and an object in motion tends to stay in motion"?

Pumping car tires

How much work does it take
to pump up a car tire?

HINT: Force times distance equals pressure times
volume.

HINT: What is the pressure of a car tire?

HINT: 10^5 N/m² equals 1 atmosphere equals 15 pounds
per square inch.

HINT: What is the volume of a car tire?

ANSWER: We defined work earlier as the product of force and distance. We could estimate the work needed to fill a car tire using a bicycle pump by estimating the force applied to the pump, the distance the pump handle travels, and the number of pump-strokes needed to fill the car tire. However, we can also do it more directly, using the pressure and volume of the car tire itself. To do this we use the fact that pressure equals force divided by area, and volume equals distance times area. This means that force times distance equals pressure times volume, or

$$W = F d = P V.$$

We'll try both methods.

It will take a few pounds of force (more than 1 and less than 30) to push the bicycle pump handle, so we will estimate 6 pounds, or 30 N. The pump handle travels about 1 foot, or 0.3 m. It will take more than 10 and less than 10^3 strokes to fill the tire, so we will estimate 10^2. This means that the total work done is

$$W_{pump} = F d = (30\,\text{N}) \times (0.3\,\text{m}) \times 100 = 10^3\,\text{J}.$$

Now let's try to estimate the pressure and volume. A full car tire is at a pressure of about 30 psi (pounds per square inch). This is 2 atmospheres, or $P = 2 \times 10^5$ N/m².[*]

To estimate the volume of a tire, we need its rim radius, thickness (sidewall height) and width. A typical compact car tire is about 8 inches (20 cm) wide with a 4-inch (10 cm) sidewall thickness, and an 8-inch

[*] There are far too many units used for pressure. One atmosphere, which is the average pressure exerted at ground level by the weight of all the air above us, equals about 15 psi, or 10^5 N/m². Less useful (but also common) units include torr (mm of mercury), inches of mercury, and inches of water, where 1 atm = 760 torr = 30 inches of mercury = 400 inches of water. Barometric pressure is commonly measured (at least in the United States) in inches of mercury.

(20 cm) radius rim. If we want more precision, we can measure these directly or just read them off the tire label.[*] A tire has a complicated shape. Let's approximate its volume by "unrolling the tire" so that it will be the product of its circumference, sidewall thickness, and width:

$$V = (2\pi r)tw = 6(20\,\text{cm})(10\,\text{cm})(20\,\text{cm})$$
$$= 2 \times 10^4\,\text{cm}^3 = 20\,\text{L}.$$

This is about 5 gallons, which seems reasonable.[†]

To calculate the work done inflating a 20-L volume, we first need to convert 20 L to 0.02 m³ (because $1\,\text{m}^3 = 10^3\,\text{L}$). The work done inflating a $0.02\,\text{m}^3$ volume to a pressure of $2 \times 10^5\,\text{N/m}^2$ is just

$$W_{\text{tire}} = PV = (2 \times 10^5\,\text{N/m}^2) \times (0.02\,\text{m}^3)$$
$$= 4 \times 10^3\,\text{J}.$$

This is four times the previous estimate. If we want to reduce the error, then we need to include the fact that the pressure increases from zero as we start inflating the tire to 30 psi when the tire is fully inflated. This means that the average pressure during inflation is only 15 psi, and we should divide our final answer by two, so that $W_{\text{tire}} = 2 \times 10^3\,\text{J}$.[‡]

As usual, we need to ask if this is a lot or a little. Inflating a car tire using a bicycle pump is certainly a significant exertion. However, 2 kJ is the energy used by a 100-W light bulb in 20 seconds and is less than 10^{-3} of a kilowatt-hour. Perhaps we can harness our cubic meter of mousetraps to do the work for us.

[*] A "P225/50R16 91S" tire has a 225-mm width, a sidewall that is 50% of the width, and a 16-inch diameter rim.

[†] If we want to calculate the volume more precisely, then we should use $V = (\pi(r + t)^2 - \pi r^2)w$. However, the approximation used above is good when the sidewall thickness, t, is significantly less than the rim radius, r.

[‡] This is the factor of one-half that we casually dismissed in the mousetrap question.

Pumping bike tires

How much work does it take
to pump up a bicycle tire?

HINT: Force times distance equals pressure times
volume.
HINT: What is the pressure of a bicycle tire?
HINT: 10^5 N/m² equals 1 atmosphere equals 15 pounds
per square inch.
HINT: What is the volume of a bicycle tire?

ANSWER: This question is identical to the previous one, except that the volume is smaller and the pressure is larger. There is also a personal difference, in that I have manually inflated many more bicycle tires than car tires.

Let's apply the two estimation techniques again. In order to estimate the work done inflating the skinny, high-pressure tires on my road bike using a bicycle pump, we need to estimate the force applied per stroke, the distance per stroke, and the number of strokes. Because the tire pressure is higher, the force applied will also be higher. The maximum force needed is easy to apply with a floor pump and difficult to apply with a hand pump, so we'll estimate that it is 40 pounds, or 200 N. The stroke length is still about 1 foot, or 30 cm. However, now it only takes about ten or twenty strokes to inflate the tire. This means that the work done is

$$W_{\text{pump}} = NFd = 20 \times (200\,\text{N}) \times (0.3\,\text{m}) = 10^3\,\text{J},$$

or a bit less than that required for the car tire.

Now let's estimate the pressure and the volume. A 100-psi bike tire is at a pressure of 7 atmospheres, or $P = 7 \times 10^5 \text{N/m}^2$. A typical road bike has a 26-inch diameter wheel with a 1.25-inch tire. Converting to sensible units (at least, units that are sensible for estimating) this is a 30-cm radius wheel with a 3-cm tire. That gives us a tire volume of

$$V = (2\pi r)tw = 6(30\,\text{cm})(3\,\text{cm})(3\,\text{cm})$$

$$= 2 \times 10^3\,\text{cm}^3 = 2\,\text{L} = 2 \times 10^{-3}\,\text{m}^3.$$

The work done inflating a 2×10^{-3} m^3 tire to $P = 7 \times 10^5$ N/m^2 is

$$W_{\text{tire}} = \frac{1}{2}PV = 0.5(7 \times 10^5\,\text{N/m}^2)$$

$$\times (2 \times 10^{-3}\,\text{m}^3) = 7 \times 10^2\,\text{J},$$

or about three times less than that required for a car tire. The volume is ten times less, but the pressure is three times larger. These two estimates agree much better because I have a lot more experience with bicycle tire inflation than I do with car tire inflation. However, even in the previous question, the estimates agreed within a factor of two.

This is the energy used by a 100-W light bulb for only 7 seconds.

Atomic bombs and confetti

5·9

Enrico Fermi estimated
the energy released by
the Trinity atomic
bomb test by
dropping
scraps of
paper and
measuring their
displacement
by the shock
wave as it passed
through them. If Fermi had
been 10 miles (16 km)
from ground zero and the
paper scraps moved 2.5 m out
(and then 2.5 m back), what
was the yield of the
Trinity test?

HINT: The work done by the shock wave is the pressure
times the change in volume.
HINT: Estimate the overpressure of the shock wave.
HINT: The change in volume is the result of the change in
radius of the sphere from 16 km to 16 km plus 2.5 m.
HINT: 1 kiloton of TNT $= 4 \times 10^{12}$ J.

ANSWER: We will estimate the energy released by an atomic bomb by using the exact same technique we just used to estimate the energy stored in a tire.[*] As the hemispherical shock wave from the atomic bomb passes, the air is displaced by a certain distance and then returns to its original location. If there were lots of pieces of confetti spread through the atmosphere, we would see a hemisphere of confetti (i.e., all of the confetti located a certain distance from the explosion) expand and then contract. As the shock wave expands, its energy will be spread over a larger hemisphere, and thus its amplitude will decrease.

The energy of the shock wave will equal the change in volume of the hemisphere as it expands (or contracts) times the pressure of the shock wave. We have the data for the hemisphere; we only need to estimate the pressure. The frontal surface area of a person is about 1 m^2. If the shock wave pressure was 1 atmosphere (10^5 N/m^2), then the force exerted on a person would have been 10^5 N or the weight of 10^4 kg (10 tons). Even 1 ton of force would have had unfortunate consequences for the observers. Therefore, the pressure was almost certainly less than 10^4 N/m^2 and more than 10^2 N/m^2 (as that would have had very small effects). We'll estimate $P = 10^3$ N/m^2.

The change in volume of the hemispherical shell is the area of the hemisphere times the displacement of the paper (its thickness):

$$\Delta V = (2\pi r^2)d = 6(1.6 \times 10^4 \, \text{m})^2 \times (2.5 \, \text{m})$$

$$= 4 \times 10^9 \, \text{m}^3.$$

[*] OK. It is not exactly the same. We will not try to estimate the bomb yield by comparing it to the work done by a bicycle pump. However, we will estimate the bomb yield by using the pressure and volume.

Thus the energy of the shock wave as it passed Fermi was

$$E = P \Delta V = (10^3 \, \text{N/m}^2) \times (4 \times 10^9 \, \text{m}^3)$$

$$= 4 \times 10^{12} \, \text{J} = 1 \, \text{kiloton},$$

using the conversion factor that $1 \, \text{kT} = 4 \times 10^{12} \, \text{J}$.[*]

Now we just need to estimate the fraction of the bomb's energy that went into the shock wave. The energy of the bomb was emitted in different forms. Some went into light (photons), some into nuclear radiation (neutrons), and some into the shock wave. In addition, the shock wave would transfer energy to the ground (by breaking things and heating things). Because there are three or four forms of emitted energy, we should multiply our estimate of 1 kT by a factor of three or four. This gives an overall estimate of 4 kT.

Based on this data, Fermi estimated that the bomb yield was 10 kT [24], which was about half of the actual yield. We did well to get within a factor of ten.

---

[*] We could have estimated that conversion factor using the energy density of gasoline, 4×10^7 J/kg, and the fact that TNT has about one-tenth the energy density of gasoline.

Energy and Transportation

Chapter 6

We use a lot of energy to move things around. We try to do this as efficiently as possible, and we worry about the relative merits of bicycles, high-speed trains, hybrid cars, electric cars, biodiesel, airplanes, solar panels, etc. Some people also worry about consuming local foods to reduce transport costs. In this chapter we will look at many of these options and try to determine the relative costs of each. Of course, we will only be able to put them into "Goldilocks" categories, but that will be good enough for most purposes.

We will look at the energy needed to move things. This will be dominated by various types of friction, especially by air resistance (or drag) and the rolling resistance of the wheels or tires on the rails or roads. There are two pieces to air resistance, the drag caused by the front of the object moving the air aside and the drag caused by friction between the sides of the object and the air it moves through (skin friction). The force exerted by the air on an object such as a truck moving through it is

$$F_{\text{drag}} = \frac{1}{2}C\rho A v^2,$$

where $\rho = 1$ kg/m^3 is the density of air, A is the area of the moving object, and v is the speed of the object. So far, it all makes sense. As the truck gets bigger (A increases), it has to move more air aside; as the truck moves faster (v increases), it has to move more air aside (the first power of v), and it has to move the air aside faster (the second power of v); if the air gets heavier (ρ increases), it is harder to move aside. C is a constant that includes all the really hard-to-calculate stuff, including how aerodynamic that truck or car really is. C for cars (which are fairly aerodynamic) ranges from about 0.25 to 0.5. C for a flat plate is 1.

Differences between 0.5 and 1 are not that important in this book (but they would make the difference between a car that gets 40 mpg and a car that gets 20 mpg). However, C for skin friction (the air resistance on the sides of the truck) is much lower, typically about 3×10^{-3}. This means that we can generally ignore skin friction except for objects that are much longer than they are wide (such as trains).

In order to keep a vehicle moving at constant speed, the engine must supply a force that exactly opposes the force of friction. The energy supplied by the engine is then the force times the distance moved. When the force is in newtons and the distance is in meters, then the energy needed will be in joules:

$$E = Fd.$$

The other major piece of information is the energy contained in gasoline. One liter of gasoline contains 3×10^7 J of chemical energy.

We also can measure energy in calories. One calorie (1 cal) equals 4 joules. However, we measure food intake in kilocalories, so 1 food calorie (1 Cal) equals 10^3 cal equals 4×10^3 joules.

Note, however, that thanks to the second law of thermodynamics,[*] the efficiency of an engine to convert the chemical energy of gasoline to mechanical energy is only about 25%. This is also true for the efficiency of our bodies to convert the chemical energy of food to mechanical energy.

Now that we have the information we need, let's get going.

---

[*] The three laws of thermodynamics are (1) you can't win, (2) you can't break even, and (3) you have to play.

Cars get between 10 and 60 miles per gallon. How many miles per gallon would a human achieve walking or running (if we could metabolize gasoline)?

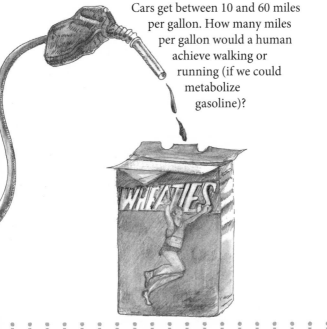

HINT: We each consume about 2,500 food calories, or 10⁷ joules, per day.

HINT: Most of that energy is devoted to maintaining our metabolisms.

ANSWER: In order to estimate our miles per gallon, we need to estimate the distance we travel and the energy needed to do so. Humans, when in reasonable shape, can walk or run long distances. We can travel more than 10 miles per day and fewer than 100, so let's take the geometric mean and estimate 30 miles. That is about the length of a marathon, so it is certainly possible for many people (although it would take me a lot more than a few hours).

In order to estimate the energy needed, we could try to analyze the energy demands of running and walking, or we could just consider the extra food needed relative to our basic metabolism. The second is easier, so we'll do that first.

In order to walk 30 miles per day, we would probably need to double our normal food intake of 10^7 J/day. Gasoline contains 3×10^7 J/L or about 10^8 J/gal. This means that we could achieve about

$$e = \frac{30 \text{ miles}}{10^7 \text{ J}} \times (10^8 \text{ J/gal}) = 300 \text{ miles/gallon}.$$

Wow. That is a lot of miles per gallon.

Alternatively, we could estimate that walking burns a couple of hundred calories per hour, so that 10 hours of walking (30 miles at 3 mph) would burn about 2,000 more calories. Remember that those are food calories, so $2 \times 10^3 \text{Cal} = 10^7$ J. This agrees quite nicely with the previous estimate.

Now let's try to directly estimate the energy needs of walking. If we walk quite briskly at 4 mph (2 m/s) we can complete our 30 miles (50 km) in about 8 hours. This means that our kinetic energy will be

$$K = \frac{1}{2}mv^2 = 0.5 \times (100 \text{ kg}) \times (2 \text{ m/s})^2 = 200 \text{ J}.$$

However, that is just the energy needed to start moving. It says nothing about the energy needed to keep moving.

At about 1 m per step, it will take about 5×10^4 steps to cover 50 km (30 mi). Walking is not 100% efficient; we will lose some of our kinetic energy at each step. We will lose more than 1% of our kinetic energy but less than 25%, so we will estimate 5%. This means that we will need to supply $E = 200\,\text{J} \times 0.05 = 10\,\text{J}$ at each step to maintain our speed. We will also need to include a 25% efficiency factor for converting food (chemical) energy into walking (kinetic) energy. Thus, the total energy needed would be

$$E = 4n_{\text{step}}E_{\text{step}} = 4(5 \times 10^4 \text{ steps})(10\,\text{J/step})$$
$$= 2 \times 10^6 \text{ J}.$$

This is a factor of five less than the previous estimates.

We can estimate this in yet another way. A small car weighs about 10^3 kg and gets about 30 mpg. Our mass is ten times less than that car, and our metabolic efficiency (excluding basal metabolism) is probably about the same as the car's (about 25%). Walking is less efficient than rolling, but air resistance is much less at walking speeds than driving speeds. Thus we should get about ten times the mileage, or about 300 miles per gallon.

Three of our four estimates are reasonably close to each other. This implies that we overestimated the efficiency of walking.

Thus, if humans could metabolize gasoline, we could achieve about 300 miles per gallon. According to David Swain, University Professor of Exercise Science at Old Dominion University, various studies have measured the energy cost of walking and determined it to be 50 kcal/mile (2×10^5 J/mile) for a 62 kg (140 lb) person. That works out to 600 mpg.

Note that fat and gasoline contain almost the same amount of chemical energy. While it is quite impressive that we can walk hundreds of miles on a single gallon of gasoline, it also means that we need to walk dozens of miles to burn off a single pound of fat.

Driving across country

How much energy does it take to transport 1 ton of cargo across the United States by automobile?

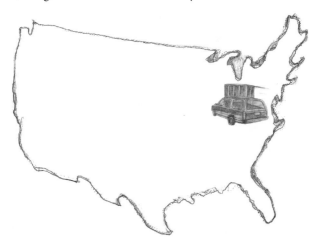

HINT: The energy efficiency is about 25%.

HINT: Most of the energy is spent overcoming air resistance.

HINT: The force of air resistance is $F_{air} = (1/2)C A\rho v^2$.

ANSWER: If there were no friction or air resistance, a car that started in New York City at 65 mph (30 m/s) would continue at that speed in a straight line forever.[*] Its mileage would be infinite, as it would have traveled thousands of miles with zero gallons of gasoline. However, cars experience both friction and air resistance. In order to estimate the energy used traveling across the country, we need to estimate these forces. Note that the force that the car's engine needs to exert is just equal to and opposite of the total force that slows it down.

The frictional force equals the force needed to push a stalled car.[†] Although it takes a lot of force to get a car started moving, once the car is moving, we can keep it moving with a force of only 10 or 20 pounds (50 to 100 N). This is about 1% of the weight of the car, implying that the coefficient of rolling resistance (tire friction) is about 1%.

Unlike friction, air resistance increases rapidly with speed. Using

$$F_{\text{air}} = \frac{1}{2} C A \rho v^2,$$

we need to estimate C, A, and v. At 65 mph, $v = 30$ m/s (because 1 m/s $\approx$ 2 mph). A car is about 6 ft (2 m) wide and 5 ft (1.5 m) tall, giving a frontal area of $A = 3 \, \text{m}^2$. Most cars are well streamlined so that $C \approx 0.25$. This gives

$$F_{\text{air}} = \frac{1}{2}(0.25)(3 \, \text{m}^2)(1 \, \text{kg/m}^3)(30 \, \text{m/s})^2 \approx 400 \, \text{N},$$

which is much more than the rolling resistance.

The distance from New York to Los Angeles is about 3,000 miles (5,000 km). Including a factor of

---

[*] Or until it hit something.

[†] If you have never needed to push a car, you are lucky (or just drive better cars).

four for engine efficiency, the energy needed to move a car from NY to LA (ignoring stoplights and mountains) is

$$E = 4F_{air}d = 4(400\,\text{N})(5 \times 10^6\,\text{m}) = 8 \times 10^9\,\text{J}.$$

This is the energy contained in about 300 L (80 gallons) of gasoline, resulting in about 40 mpg.

Thus, the energy used to drive at highway speed from NY to LA is largely due to air resistance.

However, this estimate just accounts for the energy needed to drive a car across country. It does not account for the cargo capacity. A typical medium-size car can carry about half a ton (the weight of about six people), so we would need two cars to carry an entire ton.

This means that we would need 600 liters (150 gallons) of gasoline to transport 1 ton across country using medium-size cars. If you used a Smart Car or a Hummer, your numbers would be a bit different.

Keep on trucking

How much energy does it take to transport 1 ton of cargo across the United States by truck?

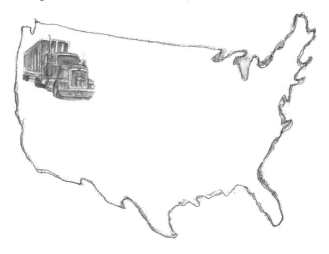

ȼ ȼ

HINT: This is very similar to the previous question.

ANSWER: This question should be easier, as we can use most of the answer to the previous question. We just need to estimate the frontal area of a semitrailer truck, its aerodynamic coefficient, its rolling resistance, and its carrying capacity.

Trucks are wider and taller than cars. A truck is about 8 feet (2.5 m) wide and less than 14 feet (4 m) tall, so it has a frontal area of about 10 m². It is definitely less aerodynamic than a car ($C = 0.25$) and more aerodynamic than a flat plate ($C = 1$), so let's estimate $C = 0.5$. This gives a force of air resistance of

$$F_{air} = \frac{1}{2}(0.5)(10 \, m^2)(1 \, kg/m^3)(30 \, m/s)^2$$

$$\approx 2 \times 10^3 \, N,$$

or about five times more than an automobile.

The rolling resistance (tire friction) will depend on the weight of the truck but not its speed. Assuming a 40-ton semitrailer truck and using the coefficient of rolling resistance of 1% that we estimated for a car, we get a force owed to the tires of 1% of the truck's weight:

$$F_{tire} = (10^{-2})(40 \, tons) = (10^{-2})(4 \times 10^5 \, N)$$

$$= 4 \times 10^3 \, N,$$

or about twice the force of air resistance.

Adding these forces and applying them over a distance of 5×10^6 m from NY to LA, requires an energy input of (including a factor of four for engine efficiency):

$$E = 4(F_{air} + F_{tire})d = 4(6 \times 10^3 \, N)(5 \times 10^6 \, m)$$

$$= 10^{11} \, J.$$

This is the energy contained in about 4×10^3 L (10^3 gallons) of gasoline or diesel fuel, which gives about 3 mpg.

Comparing our estimates to reality, we find that trucks get about 6 to 8 miles per gallon and that air resistance accounts for two-thirds of the energy used by a semitrailer traveling at 65 mph [25]. These both differ by a factor of two from our estimates, implying that we overestimated tire friction.

Now let's calculate the mileage-per-ton. The semitrailer burns about 4×10^3 liters of fuel in transporting 40 tons across country. This means that transporting 1 ton by truck across country requires about 100 L (25 gallons) of diesel fuel. That is about six times more efficient than a car.

Keep on biking

How much energy does it take to transport 1 ton of cargo across the United States by bicycle?

‿‿‿‿‿‿‿‿‿‿‿‿‿‿‿‿‿‿‿‿‿‿‿‿

HINT: This is very similar to the two previous problems.
HINT: Most of the energy is spent overcoming air resistance.
HINT: The force of air resistance is $F_{air} = (1/2)C A \rho v^2$.
HINT: How much weight can a bicyclist carry or tow?

ANSWER: Bicycles are often presented as green alternatives to automobiles for commuting relatively short distances. Here, in an effort to be utterly unreasonable, we will estimate their efficiency for hauling cargo long distances. Because we're not totally masochistic, we'll choose a road bike with narrow, high-pressure tires in order to minimize rolling resistance. That means that most of the energy will be spent fighting air resistance, and we can apply exactly the same techniques to bicycles that we just applied to cars and trucks.

The force of air resistance will be

$$F_{air} = (1/2)C A\rho v^2,$$

and the energy used to bicycle across country, including a factor of four for human metabolic efficiency, will then be

$$E = 4F_{air}d = 4(1/2)C A\rho v^2 d,$$

where C is the drag coefficient (i.e., how nonstreamlined it is), A is the frontal surface area, $\rho \approx 1\,\mathrm{kg/m}^3$ is the density of air, v is the bicycle's speed, and $d = 5 \times 10^6$ m is the distance from New York to Los Angeles.

Let's consider two extremes. If we ride a standard road bike, then we will have a relatively upright posture, so our frontal surface area will be about

$$A = 2\,\mathrm{ft} \times 5\,\mathrm{ft} = 10\,\mathrm{ft}^2 \approx 1\,\mathrm{m}^2$$

Standard road bikes are not streamlined, so our drag coefficient will be close to 1. On the other hand, we could ride a spiffy recumbent with a windshield that will reduce our frontal surface area to $A = 0.5\,\mathrm{m}^2$ and our drag coefficient to $C = 0.2$, reducing the force of air resistance by a factor of ten.

Our speed will be somewhere between 10 and 20 mph (5 and 10 m/s). After riding a few hundred miles, we'll probably be in excellent shape, so let's use 10 m/s.[*]

We will need energy of

$$E = 4F_{air}d = 4(1/2)C\rho A v^2 d$$

$$= 4(0.5)(1)(1\,kg/m^3)(1\,m^2)(10\,m/s)^2(5\times10^6\,m)$$

$$= 4(50\,N)(5\times10^6\,m)$$

$$= 10^9\,J$$

to pedal our road bike (and a lot less to pedal our spiffy recumbent).

Now we need to consider how much weight we can tow behind our bicycle. We can certainly tow more than 10 kg (20 pounds) and less than 1,000 kg (1 ton) so let's take the geometric mean and estimate 100 kg.[†] At 100 kg per bicycle, we will need ten bicycles to transport each ton, so we will need

$$E = 10^{10}\,J$$

to transport each ton across country. This is equal to 2×10^6 food calories (1 Cal = 4×10^3 J), or the energy equivalent of 300 liters of gasoline. (These numbers will be significantly less for our recumbent.)

That is a lot of food. Let's try to estimate this another way. One hundred miles in a day is a very long bicycle ride, especially towing 100 kg. If we bicycle

[*] There are trade-offs here. As we go faster, we need more energy but less time (a significant consideration when bicycling cross-country). In addition, as we go slower, air resistance becomes relatively less important, and tire friction becomes more important.

[†] This assumes that the towed weight will not significantly increase our air resistance and therefore will not significantly increase the energy expended. This will only be true if we are not towing too much weight and if the trailer is well designed. Of course, the trailer will make it much more difficult to climb hills. On the other hand, it will make going downhill much more exciting.

that far, we will probably about double our normal food intake from 2,500 to 5,000 calories per day. At 100 miles per day, it will take 30 days to bicycle across country. In that time we will consume an extra 8×10^4 calories. Ten of us will consume about an extra 10^6 calories. This is within a factor of two of our previous estimate.

Thus, bicycles are more energy efficient than cars but less energy efficient than trucks for long-distance cargo hauling.

However, labor costs change the balance completely. It takes two cars about one week each to transport 1 ton across country, whereas it will take ten bicycles about four weeks each. At $1,000 per person per week, that is $2,000 per ton for cars, $40,000 per ton for bikes, and only $25 per ton for a 40-ton truck.

How much energy does it take to transport 1 ton
of cargo across the United States by train?

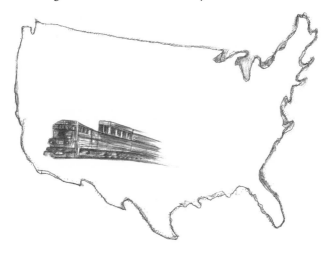

HINT: Consider the air resistance of the sides of the train.

HINT: This is very similar to the previous question.

ANSWER: This question is a little different from the two previous ones. Trains are very long compared with cars and trucks, so we will need to estimate the air drag caused by skin friction, that is, the air resistance caused by the sides and tops of the railroad cars. This will probably be more important than the frontal air resistance, especially for a very long train. In addition, steel wheels rolling on steel track have very little rolling resistance, so we will neglect that.

The frontal area of a freight train is about twice as large as that of a truck, or about 20 m^2.

Now let's consider the air resistance caused by the sides and tops of the cars. This is a very complicated problem in computational fluid dynamics. In order to solve it, first we will set up the necessary differential equations, remembering that air is a compressible fluid, and then ...

Sorry about that. Let's try again. The air resistance resulting from air flow along the sides and top of a given boxcar will be between 1% and 100% of the frontal air resistance of the train. Taking the geometric mean, we estimate that the side and top air drag of a single railroad car is 10% of the frontal air drag of the locomotive.

This means that each car will contribute the equivalent of 2 m^2 frontal area, or about one-fifth that of a semitrailer truck. The total air drag of a one hundred–car freight train will be

$$F_{\text{air}} = \frac{1}{2} C \rho A v^2$$

$$= \frac{1}{2}(0.5)(1 \,\text{kg/m}^3)(100 \times 2 \,\text{m}^2)(30 \,\text{m/s})^2$$

$$\approx 5 \times 10^4 \,\text{N},$$

where we have neglected the 20 m^2 frontal area of the locomotive because it is much smaller than the 200 m^2 total effective area of the cars.

Thus, the total energy required to move a one hundred–car freight train a distance of 5×10^6 m, including the factor of four for engine efficiency, is

$$E = 4F_{air}d = 4(5 \times 10^4 \, \text{N})(5 \times 10^6 \, \text{m}) = 10^{12} \, \text{J}.$$

This is the energy contained in 3×10^4 L (10^4 gallons) of gasoline.

Wow. That's enough to fill a small swimming pool. However, that is the wrong comparison. Note that 3×10^4 L has a mass of 3×10^4 kg, or 30 tons. That is about one-third the capacity of a single railroad tank car.

Anyway, we're supposed to be calculating the fuel needed to transport 1 ton across country. Each railroad car has a much greater capacity than a semitrailer. Many railroad cars are designed to carry two 40-foot containers, or the equivalent of two semitrailer loads. A one hundred–car freight train carrying 80 tons per car can transport 8×10^3 tons.

Thus, it takes about 4 L (1 gallon) to transport 1 ton across country by railroad. This is much more fuel efficient than other forms of transportation.

And now for the dreaded reality check. According to the American Association of Railroads, it takes about 7 gallons of fuel to transport 1 ton across country. We are within a factor of ten, but perhaps we could have done better. We neglected rolling resistance and other forms of friction. Our estimate of the air resistance was also very crude and could well have been off by a factor of several.

Keep on flying

How much energy does it take to transport 1 ton of cargo across the United States by airplane?

HINT: What is the glide ratio of an airplane? How many feet (or meters) does a gliding airplane descend for each foot (or meter) that it travels?

ANSWER: In order to figure out how much energy an airplane uses to fly across country, we could try to figure out the air resistance (drag) that the airplane encounters. Unfortunately, airplanes are far more complicated aerodynamically than railroad cars, trains, or trucks. Fortunately, there is an easier way to estimate this. When an airplane glides, it must either slow down or descend. When it glides at constant speed, it uses the energy gained by descending to offset the energy lost by air resistance. The amount it descends is typically referred to as the "glide ratio," expressed as the number of meters (or furlongs) the airplane travels horizontally as it descends 1 meter (or furlong) vertically.

Thus, if an airplane has a glide ratio of 10:1, for every 10 furlongs that it travels horizontally the engines must supply enough energy to lift the airplane 1 furlong.[*] This means that rather than estimating the aerodynamic characteristics of our cargo airplane and the speed at which it flies, we only need to estimate its glide ratio. This is much easier.

The glide ratio must be more than 1 (well-designed bricks) and less than 100 (well-designed gliders). As usual, we will take the geometric mean and estimate that the glide ratio of our cargo plane is about 10.

Thus, air resistance will cause our 1 ton of cargo flying 5×10^6 m across country to descend a total height of

$$h = 5 \times 10^6 \, \text{m}/10 = 5 \times 10^5 \, \text{m}.$$

Needless to say, without power from the engines, it won't make it all the way across country. It will take

---

[*] You may substitute your favorite unit for furlongs in the above example if you insist.

an energy input of

$$E = mgh = (10^3 \, \text{kg})(10 \, \text{m/s}^2)(5 \times 10^5 \, \text{m})$$
$$= 5 \times 10^9 \, \text{J}$$

to keep each ton of the plane at the same altitude.

However, we need to include two more factors. The first is the familiar engine efficiency of about 25%, and the second is the mass of aircraft needed per ton of cargo. Let's estimate the airplane weight as a multiple of its cargo capacity. A 1-ton automobile can carry about 0.5 ton, so its multiple would be three (gross weight is three times cargo weight). An airplane will have a higher multiple because air is intrinsically less suitable for load bearing than concrete (e.g., wings are bigger and heavier than wheels) and a lower multiple because it will be built of lighter materials. These will offset each other, so we will use three for our multiple. Applying a factor of four for engine efficiency and a factor of three for aircraft mass gives

$$E = 3 \times 4 \times (5 \times 10^9 \, \text{J}) = 5 \times 10^{10} \, \text{J}.$$

At 3×10^7 J/L of gasoline, we will need about 2×10^3 liters, or 500 gallons, of gasoline to fly a ton of cargo across country. That is about 2 tons of fuel per ton of cargo. Unsurprisingly, this is a lot more than is needed by train, truck, or even car.

Now let's compare to reality. The actual glide (lift to drag) ratio of a 747 is about 17. It burns about 11 tons of fuel per hour at cruising speed. The empty weight is about 200 tons, and the maximum takeoff weight is about 400 tons. This means that for a 6-hour transcontinental flight, it will burn about 70 tons of fuel to carry at most 130 tons of cargo or about 0.5 tons of fuel per ton of cargo. Thus our estimate is off by less than a factor of four. Most of that error is due to our underestimate of the glide ratio.

To summarize, we estimate that the fuel needed to transport 1 ton across country by various modes is as follows:

Mode	Estimated fuel (L)	Actual fuel (L)
Car	600	600
Bicycle	300	—
Truck	100	50
Train	4	30
Plane	2,000	500

To pee or not to pee

All Nippon Airways started asking passengers in 2009 to urinate before boarding the airplane. How much fuel would airlines save per passenger or per flight if passengers did this?

¿ ¿

HINT: How much mass do we urinate?

HINT: How much fuel is needed for each kilogram of mass on the flight?

ANSWER: To answer this question, we need to estimate the amount of mass excreted and the effect this would have on fuel consumption. We already estimated the volume of a typical urination to be 0.4 L when considering the amount of pee in a pool.

However, the amount of urine that an airline passenger has at the moment of boarding will vary randomly between 0 and 0.4 L. Thus, we should use 0.2 L (with a mass of 0.2 kg) for the average amount of pee. If all passengers peed before boarding (and good luck with that!), this would reduce the total takeoff mass by 80 kg on a four hundred–passenger jet. This is about the mass of one passenger.

Now we need to estimate the fuel saved. Let's consider a large jet flying 5×10^3 km (3,000 miles). We can use the total fuel needed per kilogram from the previous problem, or we can try a new way to estimate it, using the costs of the flight.

The absolute lowest fare that an airline charges should be a bit more than the extra cost of flying the extra passenger. The lowest round trip fare from NY to LA is about $250. If fuel represents half the extra cost of flying, then the additional fuel cost to fly one passenger one way is $60. At $3 per gallon, this is 20 gallons or 80 L or about 1 kg of fuel per kg of cargo.

In the previous problem we estimated that it took an average of 2 tons of fuel per ton of cargo. However, this included the fuel needed to fly the empty airplane. If we just include the extra fuel needed for each extra ton of cargo, then we would have estimated one-third of that, or about 0.6 tons of extra fuel per extra ton of cargo. This is consistent with our cost-based estimate of 1 kg of fuel per kg of cargo.[*]

Because the total weight of pee that could be excreted from a 747's passengers prior to flight is

---

[*] Or, if you prefer obscure English units, one stone of fuel per stone of cargo.

equivalent to the weight of one passenger, the airline could save a maximum of 80 L of fuel on a cross-continent flight.

Looking at this another way, it takes about a liter of extra fuel to fly 1 kg of passenger across the country. Therefore, by excreting 0.2 L of urine, each passenger will save less than 0.2 L of jet fuel. At $3/gallon, this is $0.15 each. It hardly seems worthwhile.

Now is this a cross-continent or a cross-incontinent flight?

Solar-powered cars

The Toyota Prius comes with the option of a
"Solar-Powered Ventilation System." What
fraction of a car's energy usage could a solar
panel provide?

HINT: The solar power density outside the Earth's
atmosphere is 10^3 W/m^2.
HINT: A liter of gasoline contains 3×10^7 J.

ANSWER: We should estimate both the energy that a car consumes and the energy that a solar panel can provide. A typical car burns about 500 gallons per year of gasoline (10^4 miles at 20 mpg), or about 1.5 gallons (6 liters) per day. At 3×10^7 J/L, a typical car uses 2×10^8 J of chemical energy daily.

Now we need to include the conversion efficiency from chemical to thermal to mechanical energy of about 25%. The 2×10^8 J of thermal energy used by the car each day only produces about 5×10^7 J of mechanical energy (although all that waste heat is very useful in the winter for heating the passenger compartment).

The maximum energy that a solar panel can provide is limited by the available area and by the solar flux. The available roof area is only about 1 m². Even if we cover every square centimeter of surface with solar panel in order to achieve 4 m², it will not make much difference.

The solar flux at Earth's orbit is about 10^3 W/m². However, only about half of that reaches the ground. We'll need to divide by another factor of two for nighttime and another factor of two for the fact that the light hits the solar panel at an angle. The efficiency of a typical solar panel is more than 1% and less than 100%, so we'll take the geometric mean and estimate 10%. This means that the 10^3 W/m² gives us electrical power of

$$P = \frac{(10^3 \text{ W/m}^2) \times (0.1)}{2 \times 2 \times 2} \times (1\,\text{m}^2) = 10\,\text{W}.$$

Given that there are about 10^5 s in a day, the solar panel can produce 10^6 J = 1 MJ of electrical energy daily.

This energy is only about 2% of a car's energy use. Even if we covered the entire car with solar panels, it would only increase this to 8%. We can save at least as much energy by keeping our tires properly inflated and accelerating gently.

Thus, Toyota made the right decision when it chose to use the solar panel only to run a ventilation fan when the passenger compartment gets too warm.

Put a doughnut in your tank

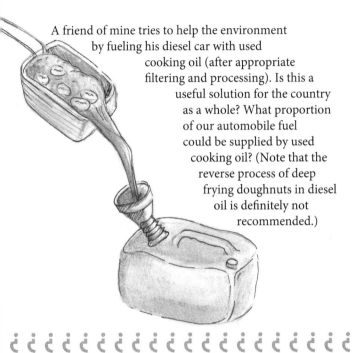

A friend of mine tries to help the environment by fueling his diesel car with used cooking oil (after appropriate filtering and processing). Is this a useful solution for the country as a whole? What proportion of our automobile fuel could be supplied by used cooking oil? (Note that the reverse process of deep frying doughnuts in diesel oil is definitely not recommended.)

HINT: How much automobile fuel do we use each day?

HINT: How much oil is used to deep fry our french fries, doughnuts, and other food each day?

ANSWER: We need to figure out how much fuel we do use and how much used cooking oil we could use. Let's start with the fuel-ish question. If you've seen this before, feel free to skip to the end of this paragraph and wait for the rest of us there. A typical American drives about 10^4 miles per year at between 10 and 40 miles per gallon. Using the geometric mean of 20 mpg, each car burns $V = 10^4$ miles/20 mpg = 500 gallons per year = 2×10^3 liters per year = 6 liters per day. With about two cars for every three Americans, that is 4 liters of automobile fuel (gasoline or diesel) per person per day.

Now let's estimate the amount of cooking oil used (but not absorbed by the food) per person per day. To do this we will need the amount of oil used per serving of deep-fried food, the number of times the oil is reused, and the number of servings consumed per day. We can safely guess that it is less than 4 liters per person.

The amount of oil needed to deep fry one serving of french fries will be more than 1 ounce and less than 1 gallon (128 oz), giving a geometric mean of 10 oz (or $\frac{1}{3}$ liter). This seems reasonable, as the french fries need to be completely immersed in the oil.[*]

One batch of cooking oil will be used to deep fry more than one and fewer than one hundred batches of french fries, so we'll take the geometric mean and estimate ten. This means that only 1 new ounce of cooking oil is used per batch of french fries.

The average American consumes more than one serving of deep-fried food per week and fewer than

---

[*] I did these estimates using U.S. customary units because that is what I use to cook with. I am therefore much more familiar with the otherwise bewildering plethora of U.S. volume units than with liters and milliliters. It is much easier for me to estimate many quantities in U.S. customary units and then convert them to metric units prior to performing calculations.

one per day. Taking the geometric mean of 0.3 servings per day, this gives 0.3 ounces (or 10 ml) of cooking oil used for deep frying per person per day. This is about 2×10^{-3} of the automobile fuel consumed. While burning used cooking oil in our cars is commendable, it will not contribute significantly to our energy needs.

Now let's compare to reality. According to Wikipedia, that oft-cited and utterly reliable source, the world consumed about 80 million tons of vegetable oil in 2000. If the United States consumed 10% of that, that amounts to 10^{10} kg per year, or 30 kg per American per year or 0.1 kg per American per day. That is ten times more than our estimate of the deep-frying oil, but it includes all uses of oil. That is still only 2% of the automobile fuel consumed.

Note: converting food crops (oil seeds or corn) to automobile fuel is a bad idea. We produce barely enough food for the world now, and cars consume far more energy than people.

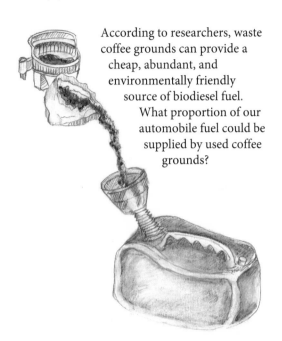

According to researchers, waste coffee grounds can provide a cheap, abundant, and environmentally friendly source of biodiesel fuel. What proportion of our automobile fuel could be supplied by used coffee grounds?

HINT: Use the automobile fuel estimate from the previous problem.
HINT: How much coffee grounds do we use per day?
HINT: What fraction of the coffee grounds is oil?

ANSWER: Some researchers claim that biodiesel derived from used coffee grounds can provide significant amounts of fuel [26].[*] In order to estimate the potential contribution of coffee grounds to our fuel consumption, we need to estimate our fuel consumption, coffee consumption, and the amount of oil we can extract from coffee grounds. In the previous problem, we estimated our average automotive fuel use at 4 liters per person per day and our average deep-frying cooking oil use at 10 ml per person per day. Now we just need to estimate the biodiesel we can make from coffee grounds.

Probably about half of all Americans drink coffee. We consume more than one and fewer than ten cups per day, so we estimate that our total daily coffee consumption is three cups per coffee drinker. It takes about one tablespoon[†] (15 mL) of coffee to make each cup, so we use

$$V_{grounds} = (3 \times 10^8 \text{ Amer})(0.5)(3 \text{ cups/day})$$
$$\times (10 \text{ mL ground coffee/cup})$$
$$= 10^{10} \text{ mL ground coffee/day}$$
$$= 10^7 \text{ L ground coffee/day},$$

or about 10^7 kg of ground coffee per day.[‡]

The amount of oil in the coffee grounds is much less well known. As usual, when we don't know what we're doing, we will bound the answer. The proportion of oil in coffee grounds must be more than 1% (to be useful as a fuel source) and less than 100%, so we will estimate 10%. This means that the 10^7 liters of coffee grounds would provide 10^6 liters of biodiesel per day. This seems like a lot.

---

[*] And it would smell much better than other fuels.

[†] Another archaic American customary volume unit.

[‡] This assumes a density of 1 kg/L, which is probably about a factor of three too high.

Comparing to reality, this estimate is consistent with the worldwide potential of 3.4×10^8 gallons per year of coffee-derived biodiesel calculated in the paper [26].

Before we can decide whether it is a lot or a little, we need to ask, "Compared to what?" In this case, we should compare this to our overall automotive fuel consumption. Let's do this per person rather than per country. The average American consumes two cups of coffee per day,* which requires 30 mL of coffee grounds. At 10% oil content, that would produce 3 mL of biodiesel per person per day.

That 3 mL is comparable to the 10 mL of used cooking oil we estimated in the previous problem. Both are a tiny fraction of the 4 L (4,000 mL) automotive fuel usage. Note that these results are not arguments against developing these fuel sources. They are, however, strong arguments against relying on these as fuel solutions.

Alas.

---

* We estimated 1.5, and I took the liberty of rounding up.

Don't slow down

How does the energy required to keep a car moving on the highway compare with its kinetic energy?

HINT: Most of the energy spent keeping the car moving is used to overcome air resistance.

HINT: Kinetic energy $K = (\frac{1}{2})mv^2$.

ANSWER: In order to answer this question, we need to estimate the kinetic energy of a car and the energy needed to keep it moving. The kinetic energy depends on the mass of the car and its velocity, $K = (1/2)mv^2$, so we will need to estimate those quantities. A typical small car has a mass of about 1 ton (10^3 kg). Highway speed is about 65 mph, which is about 30 m/s. Thus, its kinetic energy is

$$K = \frac{1}{2}mv^2 = \frac{1}{2}(10^3 \,\text{kg})(30 \,\text{m/s})^2 = 5 \times 10^5 \,\text{J}.$$

We estimated the force of air resistance on an automobile back at the beginning of this chapter. To save you the trouble of going all the way back there and then coming all the way back here, we'll just repeat the answer: $F_{air} = 400$ N. Because the energy used to overcome air resistance is $E = Fd$, this means that in 1 kilometer (10^3 m) we have to supply $E = 4 \times 10^5$ J to overcome air resistance. This is about the same amount of energy needed to accelerate the car from 0 to 30 m/s. Thus, just to keep moving at constant speed, for every kilometer the engine has to supply the same amount of energy as it took to accelerate the car from 0 to 30 m/s.

At 30 m/s, a car travels 1 km in time: $t = 10^3 \,\text{m}/(30 \,\text{m/s}) = 30$ s. It takes energy of $E = 5 \times 10^5$ J to accelerate a car from zero to 30 m/s (65 mph). That amount of energy needs to be replaced every 30 s to keep the car going at 30 m/s.

Note that heavier cars require more energy to accelerate because they have more mass, and bigger cars require more energy to overcome air resistance because they have more frontal area.

Throwing tomatoes

Locavores claim that we should eat food that is locally produced, not shipped from far away. How much energy does it take to ship a tomato across the country? How much energy would it take to ship one year's worth of a person's groceries across the United States?

HINT: How much does a tomato weigh?

HINT: Tomatoes are shipped by truck.

HINT: What is the mass of the groceries that we consume each year?

ANSWER: We estimated previously that it takes 100 L of fuel to transport 1 ton of cargo across country by truck. (We estimated that a 40-ton truck averaged 3 miles per gallon driving 3,000 miles.) The average tomato weighs a trifle less than 1 ton, so we need to estimate its weight. Tomatoes range tremendously in volume—from less than one ounce to about a pint. Because tomatoes have the density of water, that means that they range in weight from less than 1 ounce to about 1 pound (16 oz). Let's use a large tomato with a weight of about one pound (0.5 kg or 1/2,000 ton). Given that it takes 100 L of fuel to transport a ton, it will take

$$V_{\text{fuel}}^{\text{tomato}} = \left(10^2 \, \frac{\text{L}}{\text{ton}}\right) (0.5 \, \text{kg}) \left(10^{-3} \frac{\text{ton}}{\text{kg}}\right)$$

$$= 5 \times 10^{-2} \, \text{L} = 50 \, \text{mL}$$

of fuel to transport that tomato from California to New York (or from New York to California). That is about 2 ounces. However, we eat more than one tomato a year.

If all of our food had to be transported across country, it would take a bit more fuel. We each consume between 1 and 10 pounds of food per day. The geometric mean of 3 pounds per day multiplied by 300 days per year equals 1,000 pounds (0.5 ton) per year. At 100 L/ton, that would require

$$V_{\text{fuel}}^{\text{food}} = (10^2 \, \text{L/ton})(0.5 \, \text{ton}) = 50 \, \text{L}$$

(10 gallons) of fuel to transport our food across country.

Because we typically each consume 500 gallons of fuel each year just for our cars and SUVs, an additional 10 gallons per year to transport our food is negligible. Thanks to cheap shipping, we can enjoy all sorts of out-of-season fruits and vegetables year-round, thus avoiding pellagra, scurvy, beriberi, and other nasty deficiency diseases.

Of course, we can still consume local produce because it tastes better or because we like our local farms. We just should not feel guilty about eating food from far away.

Heavenly Bodies

Chapter 7

ن ن

Heavenly bodies attract each other.[*] They also move in circles, rotating on their axes and revolving around each other.[†] In order to describe their interaction and motion, we need to introduce gravity, tides, and rotational motion. As a result, this chapter will be the most mathematical. Because rotational motion is complicated, it will be introduced at the end of the chapter.

We'll consider two effects of gravity, force and potential energy, and then discuss tidal forces. When we are close to the surface of a planet, the gravitational force on us from the planet is approximately constant ($F_{grav} = mg$, where m is your mass and $g = 10$ N/kg $= 10$ m/s^2 is the value of the constant near the surface of the Earth), and the gravitational potential energy is simply the force times the height ($PE = mgh$). However, the further we move away from the planet, the weaker the gravitational force becomes. The force is

$$F_{\text{grav}} = \frac{GMm}{d^2},$$

where M is the mass of the planet, m is the mass of the other object (the Moon, a satellite, you, me), d is the distance from the center of the planet to the center of the other object, and G is a universal constant: $G = 7 \times 10^{-11}$ N-m^2/kg^2. Thus, if the distance from us to the center of the Earth doubles, then the gravitational force on us will drop to one-quarter as much.

According to Newton's First Law, an object does not change its speed or direction of motion unless a force acts on it. This means that satellites travel in circular orbits because of gravitational forces.

---

[*] No, this does not refer to the celebrity of the moment.

[†] In this book, almost all orbits are approximated as circles. Sorry about that, Kepler.

The force needed to make something travel a circular path is

$$F_{circ} = \frac{mv^2}{R},$$

where m is the mass moving in a circle, v is its speed, and r is the radius of the circle.[*] This applies to the Moon orbiting the Earth or to a car rounding a curve.

Because the gravitational force varies, the gravitational potential energy is no longer simply force times distance. Instead,

$$PE_g = -\frac{GMm}{d}.$$

This is defined so that our gravitational potential energy is zero when we are far from the planet and it decreases as we get closer. An object falling toward the earth will convert potential energy to kinetic energy; its gravitational potential energy will decrease, and it will speed up.

Tidal forces occur when, for example, the Moon pulls harder on the near side of the Earth than it does on the center of the earth. Tidal forces are the difference between those two forces. The difference in the gravitational force on a 1 kg object at the center of the object and a 1 kg object on its surface is

$$F_{tide} = \frac{2GM(1\ kg)r}{d^3}.$$

In the case of tides caused by the Moon on the Earth, M is the mass of the Moon, the (1 kg) refers to test masses at the center and surface of the Earth, r is the radius of the Earth, and d is the distance from the center of the Earth to the center of the Moon. This tidal force pulls objects at the surface of the Earth away from its center.

---

[*] This force points toward the center of the circle. The technical term for it is "centripetal force." It is frequently referred to colloquially (and incorrectly) as "centrifugal force."

How fast does the Earth go (i.e., what is its speed) as it travels around the Sun?

HINT: One year = $\pi \times 10^7$ seconds.

HINT: The Earth is 1.5×10^8 km (93 million miles) from the Sun.

ANSWER: In order to calculate speed, we need to estimate time and distance. The Earth goes around the Sun once a year. The Earth travels around the Sun in a circle with a radius of 1 astronomical unit (1 AU = 93 million miles = 1.5×10^8 km). The circumference of a circle is $c = 2\pi r$. Therefore the Earth's speed is $v = 6$ AU/yr. While true, this is remarkably unuseful. We need to convert the numbers to more standard units.

If we multiply the number of seconds in a minute times the number of minutes in an hour times ... then we will find that 1 year = 3.15×10^7 s $\approx \pi \times 10^7$ s.* Needless to say, the important number in that equation is the exponent 7, not the prefactor π.

If the radius of the circle is $r = 1.5 \times 10^8$ km, then the circumference is $c = 2\pi r = 10^9$ km = 10^{12} m. This means that the speed of the Earth in its orbit is

$$v = \frac{d}{t} = \frac{10^{12}\,\text{km}}{\pi \times 10^7\,\text{s}} = 3 \times 10^4\,\text{m/s} = 30\,\text{km/s}.$$

Wow. That is pretty fast.

(This question is repeated from *Guesstimation*.)

---

* I like telling my students that the constant π appears because the earth goes around the Sun in a circle and that the actual value is not quite equal to π because the orbit is not quite a circle. The ones who nod their heads dutifully at this nonsense are clearly the ones that will always vote for <insert name of hated political party here>.

Flying off the Earth

7.2

What is the shortest day the Earth could have without flying apart?

¿ ¿

HINT: The radius of the Earth is 6×10^6 m.

HINT: Gravity provides the centripetal force to keep the Earth together.

HINT: The shorter the day, the faster the rotational speed at the equator.

ANSWER: Gravity provides the force needed to keep a rock on the equator traveling in a circle as the Earth rotates on its axis. Let's choose a green rock to make the arithmetic easier.* The gravitational force on that rock is $F = mg$. The force needed to keep it moving in a circle is $F_{\text{circ}} = mv^2/r$, where v is the speed of the rock and r is the distance from the rock to the center of the circle (i.e., the center of the Earth, $r = 6 \times 10^6$ m). We can rearrange this so that[†]

$$v_{\text{max}} = \sqrt{\frac{Fr}{m}} = \sqrt{\frac{mgr}{m}} = \sqrt{gr}$$

$$= \left[(10\,\text{m/s}^2)(6 \times 10^6\,\text{m})\right]^{1/2} = 10^4\,\text{m/s}.$$

That seems rather fast. (Note that this result does not depend on the mass [or color] of the chosen rock.)

Let's compare it to the actual speed of a rock at the equator. That rock will travel completely around the circumference of the Earth every day. This means that its speed is

$$v = \frac{d}{t} = \frac{2\pi r}{1\,\text{day}}.$$

The number of seconds in one day is a very useful number to remember. There are 24 hours per day and 60 minutes per hour which gives $24 \times 60 \approx$ 1,500 minutes in a day. Multiplying by 60 seconds per minute we get $1,500 \times 60 \approx 10^5$ seconds per day.[‡]

---

* Or we could choose a 1 kg rock to make it more colorful.

[†] Fortunately, square roots are not that hard to approximate. Just divide the exponent by two and take the approximate square root of the coefficient. In this case we want the square root of 6×10^7. Because the exponent is odd, it is a little more complicated. However, here we'll just round up. This means that $\sqrt{6 \times 10^7} \approx \sqrt{10^8} = 10^4$. This is within 20% of the actual answer, which is quite good enough.

[‡] There are actually 86,400 seconds per day, but it is much easier and more useful to remember 10^5. When we need the precision of 86,400, we will also need a calculator.

The speed of our rock is then

$$v = \frac{6 \times 6 \times 10^6 \, \text{m}}{10^5 \, \text{s}} = 400 \, \text{m/s},$$

or about 10^3 miles per hour.

This means that the maximum speed at which gravity would barely keep our rock moving in a circle, $v_{\text{max}} = 10^4$ m/s, is about twenty times faster than the actual speed, $v = 400$ m/s. Thus the Earth could rotate twenty times faster before disintegrating. At that speed the day would be twenty times shorter, or about 1 hour long. The good news is that the work week would be over in under five hours. The bad news is that even a long weekend would last only 3 hours.

The rings of Earth

7·3

What is the minimum distance between the
Earth and the Moon before tides tear the
Moon apart (and it becomes a very pretty
set of rings)?

¿ ¿

HINT: The tidal force exerted by the Earth on the Moon
pulls the Moon apart. The Moon's gravity pulls it together.

HINT: The Earth's mass is $M_E = 6 \times 10^{24}$ kg.

HINT: The Moon's mass is about 1% of the Earth's.

HINT: The distance from the Earth to the Moon is
4×10^5 km.

HINT: The Moon has the same apparent size as a finger's
width at arm's length.

ANSWER: The Moon is moving away from the Earth at about 4 cm per year, owing to tidal interactions.[*] Therefore, in the distant past the Moon was significantly closer to the Earth. We want to estimate the minimum distance at which the Moon could orbit the Earth without being torn apart by tides. This is referred to as Roche's Limit. The Earth exerts a greater gravitational force on the closest part of the Moon than on the center of the Moon. If the difference in the Earth's gravitational force on those two points is greater than the Moon's gravitational force, then the Moon would be ripped apart, and the Earth would have rings like Saturn.

In order to estimate this, we need a lot of information. The tidal force of the Earth on the Moon depends on the Earth's mass, the distance from the Earth to the Moon, and the radius of the Moon. The gravitational force of the Moon exerted on a surface rock depends on the Moon's mass and radius.

Let's first look at the gravitational pull exerted by the Moon on a 1 kg rock on its surface. If you remember that the gravitational force on the Moon's surface is one-sixth that of the Earth, then feel free to skip the next couple of paragraphs. Otherwise, we need to estimate

$$F_{\text{Moon}} = \frac{G M_M (1 \text{ kg})}{r_M^2},$$

so we need to know the Moon's mass and radius. Fortunately, the size of the Moon is not hard to estimate. If you hold up a finger at arm's length, it just covers the Moon. This means that the diameter of the Moon divided by the distance to the Moon equals the diameter of your finger (1 cm) divided by the distance

---

[*] More precisely, the average distance between the Earth and the Moon is increasing by that amount. Precision of language is important; numerical precision is only sometimes important.

to your finger (1 m), or in other words, that the diameter of the Moon is 1% of the distance to the Moon.[*] Given that the distance from the Earth to the Moon is $d_{EM} = 4 \times 10^5$ km,[†] the radius of the Moon is half of its diameter:

$$r_M = \frac{1}{2}d_M = 0.5 \times (4 \times 10^8 \text{ m}) \times \frac{1 \text{ cm}}{1 \text{ m}} = 2 \times 10^6 \text{ m},$$

or 2×10^3 km. That is about one-third of the Earth's radius.

We can estimate the mass of the Moon from its size. Its radius is one-third of the Earth's, so its mass should be about $\left(\frac{1}{3}\right)^3 \approx \frac{1}{30}$ of the Earth's. In reality the Moon is significantly less dense than the Earth, so its mass is about 1% of the Earth's. Because the mass of the Earth is $M_E = 6 \times 10^{24}$ kg, the mass of the Moon is $M_M \approx 6 \times 10^{22}$ kg.[‡]

Thus, the gravitational force of the Moon on a 1-kg rock on its surface is

$$F = \frac{G M_M (1 \text{ kg})}{r_M^2}$$

$$= \frac{(7 \times 10^{-11} \text{ N-m}^2/\text{kg}^2)(6 \times 10^{22} \text{ kg})(1 \text{ kg})}{(2 \times 10^6 \text{ m})^2}$$

$$= 1 \text{ N}.$$

If you remembered that the Moon's gravity is one-sixth of the Earth's, then you got $F = (1/6)(10 \text{ N}) = 1.6$ N, or about 60% more than our estimate.

Now let's look at the tidal force that the Earth exerts on the Moon, the difference between the gravitational

---

[*] This is also true of the Sun, which is the same apparent size as the Moon.

[†] This is another one of those numbers that it is really useful to know, at least for astronomical estimation.

[‡] Yes, you can search the Internet and find out that this value is about 25% too low. If you're worried about that level of precision, why are you still reading this book?

force exerted by the Earth on a 1-kg rock at the center of the Moon, and a 1-kg rock on the near side of the Moon. This difference is

$$F_{\text{tide}} = \frac{2G\,M_E(1\,\text{kg})r_M}{d_{EM}^3}.$$

We want to set this equal to the 1 N that the Moon's gravity exerts on that rock. This means that, solving for the distance,

$$d_{EM} = \left(\frac{2G\,M_E(1\,\text{kg})r_M}{1\,\text{N}}\right)^{1/3}$$

$$= \left(\frac{2(7 \times 10^{-11}\,\text{N-m}^2/\text{kg}^2)(6 \times 10^{24}\,\text{kg})}{1\,\text{N}} \times \frac{(1\,\text{kg})(2 \times 10^6\,\text{m})}{}\right)^{1/3}$$

$$= (2 \times 10^{21}\,\text{m}^3)^{1/3}$$

$$\approx 10^7\,\text{m}.$$

Although cube roots look scary, they are not really. All we need to do is to divide the exponent by three. In this case, the exponent is 21, so that the answer is a little more than 10^7. Also, because we took the cube root, it does not matter numerically whether we used 2 N or 1 N for the gravitational force of the Moon on a 1-kg rock.

This means that the Moon could never have been closer to the Earth than 10^7 m. At this point, the distance from the surface of the Earth to the surface of the Moon would be only

$$d = 10^7\,\text{m} - 2 \times 10^6\,\text{m} - 6 \times 10^6\,\text{m} = 2 \times 10^6\,\text{m}.$$

In other words, the separation between the Earth and the Moon would be about equal to the radius of the Moon and much less than the radius of the Earth.

Compare the relative strengths of the tidal forces on a newborn baby from Saturn, the Moon, and the obstetrician.

ɔ ɔ

HINT: Saturn is about ten times further from the Sun than the Earth is.

HINT: Saturn's mass is about one hundred times the Earth's mass.

ANSWER: Astrology claims that the positions of the planets when we are born influences our personalities and our destinies. The only physical force that the planets can exert on us is gravitational. The average force will not affect us, but the difference between the forces exerted on our heads and on our feet just might. That difference is the tidal force, so let's estimate that for different objects. The object exerting the biggest tidal force is probably the Earth, but that will have the same effect on all of us, so it cannot differentiate our destinies.[*]

The tidal force is the difference in the gravitational force exerted on, say, our head and our feet. Let's consider two small pieces of us, each of mass m so that the tidal force is

$$F_{\text{tide}} = \frac{2GMmh}{d^3},$$

where M is the mass of the object attracting us, h is our height, and d is the distance from us to the center of that object. Because m and h are characteristics of us, they will be the same for all attractors. Similarly, G and 2 never change. Then the relative size of the tidal force depends only on the mass and distance of the attractor

$$F_{\text{tide}} \propto \frac{M}{d^3},$$

so we only need to estimate those masses and distances.

If we don't know the mass of Saturn, we can estimate it with respect to the Sun. The mass of the Sun is about 10^6 times greater than the mass of the Earth. The mass of the gas giant planets will be geometrically closer to the Earth's mass than to the Sun's mass. Because the geometric mean of the Earth's and Sun's

---

[*] It cannot integrate them either.

masses is 10^3 times that of the Earth, we will estimate the mass of Saturn as 10^2 times that of the Earth, or $M_S = 6 \times 10^{26}$ kg.[*]

We can estimate the distance from the Earth to Saturn in a couple of different ways. Pluto is about 100 AU from the Sun. Because there are two planets between the Earth and Saturn and two more between Saturn and Pluto, we'll take the geometric mean of 1 and 100 AU to estimate 10 AU for the distance to Saturn. Alternatively, we can estimate that the distance from the Sun to each succeeding planet doubles. This would give a distance to Saturn of $2^3 = 8$ AU. As the distance from the Sun to Saturn is much greater than the distance from the Sun to the Earth, the relative orbital positions of the Earth and Saturn can be ignored (i.e., it does not make much difference whether the distance from the Earth to Saturn is $10 + 1 = 11$ AU or $10 - 1 = 9$ AU).

We remember the mass and distance to the Moon from the previous problems as $M_M = 6 \times 10^{22}$ kg and $d_M = 4 \times 10^8$ m. We can estimate the mass and distance to the obstetrician as $M_O = 10^2$ kg and $d_O = 1$ m (or even closer).

Now it's just arithmetic:

$$F_{\text{tide}} \propto \frac{M}{d^3}$$

$$F_{\text{tide}}^{\text{Saturn}} \propto \frac{6 \times 10^{26} \text{ kg}}{(1.5 \times 10^{13} \text{ m})^3} = 2 \times 10^{-13} \text{ kg/m}^3$$

$$F_{\text{tide}}^{\text{Moon}} \propto \frac{6 \times 10^{22} \text{ kg}}{(4 \times 10^8 \text{ m})^3} = 10^{-3} \text{ kg/m}^3$$

---

[*] We got lucky here. The gas giant masses range from fourteen to three hundred times that of the Earth. Saturn's mass just happens to be almost exactly one hundred times that of the Earth's. However, it really does not matter. Our mass estimate is within a factor of ten for all four gas giants.

$$F_{\text{tide}}^{\text{obs}} \propto \frac{100 \text{ kg}}{(1 \text{ m})^3} = 10^2 \text{ kg/m}^3$$

The units are irrelevant. I include them here just to be pedantic. The important point to notice is that the tidal force on a newborn baby from the obstetrician is almost a million times larger than that of the Moon, which is a billion times larger than that of Saturn.

Why would anyone think that the planets could influence our destinies?

How closely could we safely orbit a neutron star, considering only gravitational effects?

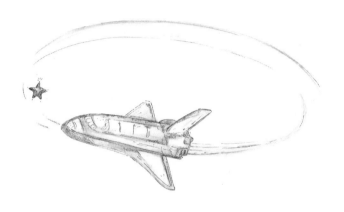

HINT: We are in free fall in orbit.

HINT: The neutron star will pull harder on the near side of our bodies than on the far side. This tidal force will stretch us.

HINT: The neutron star mass will be similar to the mass of the Sun, $M_\odot = 2 \times 10^{30}$ kg.

HINT: $F_{tide} = 2GM(\text{1 kg})r/d^3$.

ANSWER: When we orbit close to a neutron star, the immense gravitational force of the star provides the centripetal force needed to keep us moving in our orbit.

Let's assume that the neutron star has the same mass as our Sun. We can estimate this in several ways. We might remember that the Sun's mass is a million times that of the Earth, and $M_E = 6 \times 10^{24}$ kg. We can also calculate the Sun's mass from the gravitational force exerted on the Earth, $F_g = GM_\odot m_{earth}/r^2$, which provides the centripetal force for the Earth to move in its orbit, $F_c = m_{earth} v^2/r$, where r is the distance from the Earth to the Sun and $v = 2\pi(1\,\mathrm{AU})/(1\,\mathrm{yr})$ is the speed of the Earth in its orbit. That will give $M_\odot = 2 \times 10^{30}$ kg.

If we are in a circular orbit 100 km (10^5 m) from the center of a neutron star with the same mass as our sun, then the average gravitational force on us will be enormous

$$F_g = \frac{GM_\odot m}{r^2}$$

$$= \frac{(7 \times 10^{-11}\,\mathrm{N\text{-}m}^2/\mathrm{kg}^2)(2 \times 10^{30}\,\mathrm{kg})(100\,\mathrm{kg})}{(10^5\,\mathrm{m})^2}$$

$$= 10^{12}\,\mathrm{N}.$$

This gives an average gravitational acceleration of $a = (10^{12}\,\mathrm{N})/(100\,\mathrm{kg}) = 10^{10}\,\mathrm{m/s}^2$. However, as we will be in free fall, we will not feel this average force.[*] As we fall, we feel weightless, whether we are jumping off a chair, riding in the "vomit comet," or orbiting the Earth.

---

[*] When we are in orbit around the Earth or another object, we are forever falling. However, we are moving fast enough that we fall around the Earth, rather than falling to the Earth. This fulfills Douglas Adams's definition of flying as "Throwing yourself at the ground and missing" [4].

The centers of our bodies will be exactly 10^5 m from the neutron star and so will experience this force and will orbit the neutron star at a certain speed. Our feet will be about 1 m closer to the neutron star and so will experience a greater force and will want to orbit the neutron star at a greater speed. Normally, we would ignore the difference between the force exerted at $d = 10^5$ m and the force exerted at $d = 10^5 - 1$ m. However, in this case the difference of those two forces acting on two parts of our bodies acts to pull our bodies apart.

In order to estimate the minimum safe distance, we need to estimate the maximum difference in gravitational acceleration that we can tolerate. If we sit on a high surface with our lower legs dangling over the edge, we can imagine a heavy weight hanging from one foot. We can easily tolerate a 10-kg (20-lb) mass, probably tolerate a 100-kg (200-lb) mass, and definitely not tolerate a 10^3-kg (1-ton) mass. Thus, the maximum tidal force we can withstand is $F_{\text{tide}}^{\max} = mg = 10^3$ N.

The tidal force (difference in gravitational forces) on two 1-kg masses is

$$F_{\text{tide}} = \frac{2GM(1 \text{ kg})r}{d^3} < F_{\text{tide}}^{\max} = 10^3 \text{ N},$$

so the minimum safe distance is

$$d = \left(\frac{2GM(1 \text{ kg})r}{F_{\text{tide}}^{\max}} \right)^{1/3}.$$

We have already determined the distance from our centers to our feet (about 1 m) and the maximum force difference we can withstand, so now it's just arithmetic:

$$d = \left(\frac{2G\,M(1\ \text{kg})r}{F_{\text{tide}}^{\max}} \right)^{1/3}$$

$$= \left(\frac{2(7 \times 10^{-11}\ \text{N-m}^2/\text{kg}^2)(2 \times 10^{30}\ \text{kg})}{10^3\ \text{N}} \right)^{1/3}$$

$$= (3 \times 10^{17}\ \text{m}^3)^{1/3}$$

$$= 10^6\ \text{m},$$

where we take the cube root by dividing the exponent by three (and rounding up).[*]

Thus, even at a distance of 10^6 m, or 10^3 km, from a one-solar-mass neutron star, the tidal forces would be gigantic. In practice, I would want to stay at least ten times further away. The tidal forces at 10^4 km would be 10^3 times smaller, but still definitely noticeable.[†]

---

[*] We could be more precise by rewriting 3×10^{17} as 300×10^{15}. Then the cube root of 10^{15} is 10^5 and the cube root of 300 is between 6 and 7. This would change our answer from 10^6 m to 7×10^5 m. Big deal.

[†] For a great fictional treatment, see Larry Niven's short story "Neutron Star."

How high can we jump?

What is the largest moon (or asteroid) that we can jump off? In other words, assuming reasonable densities, what is the radius of the largest rocky sphere from which we can reach escape velocity by jumping?

HINT: We can jump off a moon when our initial kinetic energy exceeds our initial gravitational potential energy,
$$PE_g = -GMm/r.$$

HINT: When we jump on Earth, our initial kinetic energy equals our potential energy at the height of our jump:
$$PE = mgh.$$

HINT: The mass of a moon or asteroid increases as the cube of its radius, $M = 4\rho r^3$, where ρ is the density and $\pi = 3$.

HINT: The density of rock is greater than that of water ($\rho = 1$ ton/m^3) and less than that of iron ($\rho = 10$ ton/m^3).

ANSWER: There is no way we can jump off the Earth. Our jumping speed is on the order of m/s, and escape velocity is on the order of 10 km/s. In order to jump off of an asteroid or moon, we need to provide enough kinetic energy to overcome our gravitational potential energy. The kinetic energy of our jump will not depend on the asteroid's size; our gravitational potential energy will.

Let's start by estimating our jump energy. When we jump on the Earth, we start with a lot of kinetic energy ($KE = \left(\frac{1}{2}\right) mv^2$), convert it all to potential energy ($PE = mgh$) at apogee,[*] and then convert it back to kinetic energy as we fall back to the Earth. A typical human can jump about 1 m (more than 10 cm and definitely less than 10 m). Thus, a typical 100 kg human can jump with

$$E = (100 \text{ kg})(10 \text{ m/s}^2)(1 \text{ m}) = 10^3 \text{ J}$$

of kinetic energy.[†]

Let's now consider a moon (or asteroid). As we increase the moon's radius, its mass will also increase: $M = \rho V$, where ρ is its density and V is its volume. Its density will be between that of water ($\rho = 10^3 \text{ kg/m}^3$) and iron ($\rho = 10 \times 10^3 \text{ kg/m}^3$), so we will use the geometric mean of $\rho = 3 \times 10^3 \text{ kg/m}^3$. Then the mass of the asteroid will be

$$M = \frac{4}{3}\pi r^3 \rho = 4r^3 \rho.$$

---

[*] A fancy term for the top of the jump.

[†] Note that this value is proportional to our mass. Our final result will not depend on our mass, however, as gravitational potential energy is also proportional to mass.

Our gravitational potential energy as we stand on its surface will be

$$PE_g = -\frac{GMm}{r}$$

$$= -\frac{G4R^3\rho m}{r} = -4G\rho r^2 m.$$

In order to be able to jump off that moon, our initial kinetic energy must be greater than the magnitude of our potential energy:

$$4G\rho r^2 m < mgh$$

$$r^2 < \frac{gh}{4G\rho}$$

$$r < \left(\frac{(10\,\text{m/s}^2)(1\,\text{m})}{4(7 \times 10^{-11}\,\text{N-m}^2/\text{kg}^2)}\right)^{1/2}$$
$$\times(3 \times 10^3\,\text{kg/m}^3)$$

$$< \left(10^7\,\text{m}^2\right)^{1/2}$$

$$< 3 \times 10^3\,\text{m}.$$

(Note that $\sqrt{10^7} = \sqrt{10 \times 10^6} = 3 \times 10^3$.) Thus, we can achieve escape velocity by jumping from the surface of a moon (or asteroid) that has a radius less than 3 km. While this information is not particularly useful today, it could be quite important when we start mining the asteroid belt.

On an object this small, we would weigh only

$$F = \frac{GMm}{r^2} = \frac{G4\rho r^3 m}{r^2}$$

$$= 4G\rho r m$$

$$= 4(7 \times 10^{-11} \text{ N-m}^2/\text{kg}^2)(3 \times 10^3 \text{ kg/m}^3)$$

$$\times (3 \times 10^3 \text{ m})(100 \text{ kg})$$

$$= 0.3 \text{ N},$$

or about 1 ounce.

Note that it takes half as much energy to orbit the asteroid as to escape from it completely. If we could jump horizontally, we could easily go into orbit.[*]

[*] There is a wonderful science fiction story by Ben Bova titled "Men of Good Will" about the consequences of firing weapons on the Moon when the projectiles go into orbit.

How much energy would be released
by stellar collapse? If the Sun
collapsed into a neutron star,
how much gravitational
energy would it
release?

HINT: The mass of the Sun is $M_\odot = 2 \times 10^{30}$ kg.

HINT: The density of the atomic nucleus is
$\rho = 3 \times 10^{17}$ kg/m³.

HINT: The gravitational potential energy of two masses,
M and m, separated by a distance r is $PE = -GMm/r$.

HINT: We can approximate the gravitational potential
energy of a sphere of mass M and radius R as the energy
of two smaller objects of mass $M/2$ separated by a
distance r.

ANSWER: When very massive stars run out of nuclear fuel, they collapse. This collapse releases tremendous amounts of gravitational energy, resulting in an explosion called a supernova. If the original star is massive enough, but not too massive, then the stellar remnant becomes a neutron star, with a density about equal to that of the atomic nucleus.[*] Although the Sun is not massive enough to either cause a supernova or end up as a neutron star, it is about the same mass as a neutron star (the remnant of the explosion of more massive stars), so the gravitational energy released would be comparable.

In order to estimate the gravitational energy released by stellar collapse, we need to estimate the gravitational potential energy before and after the collapse. Our formula for gravitational potential energy,

$$PE = -\frac{GMm}{r},$$

applies to two objects of masses, M and m, separated by a distance r. We can approximate the potential energy of a spherical object by dividing it into smaller spheres and calculating the energy of each pair of spheres. The first step is to divide it into two equal mass spheres separated by the radius, r, of the initial object.[†] Thus we need to estimate the mass, current radius of the Sun, and the radius of the Sun as a neutron star.

We estimated the mass of the Sun in a previous problem as $M_\odot = 2 \times 10^{30}$ kg. The diameter of the Sun is easier. We might remember that the Sun's radius is one hundred times larger than the Earth's.

---

[*] If the original star is too massive, then the stellar remnant becomes a black hole.

[†] Yes. We could be more precise by calculating the radius of each of those smaller spheres and using that. But we won't. Well, I won't.

Alternatively, if we hold a finger (preferably our own) at arm's length, it almost exactly covers the Sun. This means that the diameter of the Sun divided by its distance equals the width of our finger divided by its distance or that

$$r_\odot = \frac{d_\odot}{2} = \frac{d_{\text{finger}}}{2l_{\text{arm}}} \times (1 \text{ AU})$$

$$= \frac{10^{-2} \text{ m}}{2 \text{ m}} (1.5 \times 10^{11} \text{ m})$$

$$= 10^9 \text{ m},$$

where we use 1 cm for a finger's width and 1 m for an arm's length.

We can determine the change in the Sun's radius from the difference between the radius of an atom and the radius of its nucleus. The size of an atom is about 1 angstrom (10^{-10} m), and the size of the atomic nucleus is about 1 fermi (1 fm = 1 femtometer = 10^{-15} m). If the constituents of the Sun decrease in radius by a factor of 10^5, then the Sun will decrease in radius by the same factor from $r_\odot = 10^9$ m to $r_{ns} = 10^4$ m. Alternatively, if we know the density of the nucleus, $\rho = 3 \times 10^{17} \text{ kg/m}^3$, then we can determine the neutron star's radius from its density and mass, $M = V\rho = 4\pi r^3 \rho$, so

$$r_{ns} = \left(\frac{M_\odot}{4\pi\rho} \right)^{1/3}$$

$$= \left(\frac{2 \times 10^{30} \text{ kg}}{12 \times (3 \times 10^{17} \text{ kg/m}^3)} \right)^{1/3}$$

$$= (5 \times 10^{11} \text{ m}^3)^{1/3}$$

$$= 10^4 \text{ m},$$

or 10 km.

Now we can plug in the numbers.

$$PE_\odot = -\frac{G(M_\odot/2)^2}{r_\odot}$$

$$= -\frac{(7 \times 10^{-11} \text{ N-m}^2/\text{kg}^2)(10^{30} \text{ kg})^2}{10^9 \text{ m}}$$

$$= -7 \times 10^{40} \text{ J}.$$

Note that this number by itself means almost nothing.[*] We need to calculate the post-collapse potential energy and take the difference between the two to determine the energy released.

Fortunately, the post-collapse potential energy is easy to calculate. Because the radius decreases by a factor of 10^5, the gravitational potential energy changes by a factor of 10^5, so

$$PE_{ns} = 10^5 PE_\odot = -7 \times 10^{45} \text{ J},$$

and the gravitational potential energy decreases by about 10^{46} J. This immense energy is transformed into other forms of energy.

This is, of course, an immense amount of energy on any human scale. But we need to compare it to cosmic scales. The Sun emits about 4×10^{26} W. There are about 10^{11} stars in our galaxy, so the galaxy emits about 10^{38} W (rounding up). Thus, 10^{46} J represents the total output of the entire Milky Way galaxy for

$$t = (10^{46} \text{ J})/(10^{38} \text{ W}) = 10^8 \text{ s},$$

or about three years! This is an immense amount of energy, even on a cosmic scale. This is the energy that powers supernovas, which outshine entire galaxies.

Wow.

---

[*] This is approximately the gravitational energy released when the Sun formed out of gas in the primordial solar system (i.e., when the gas that forms the Sun moved from very large distances where $PE = 0$ to distances within the radius of the current Sun).

Splitting the Moon

7.8

How large a collision would it take to split the Moon? In other words, how much energy would it take to split the Moon into two approximately equal pieces and separate those pieces?

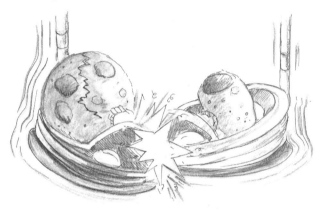

HINT: The Moon is held together by gravity.

HINT: We can approximate the gravitational potential energy of a sphere of mass M and radius r as the energy of two smaller objects of mass $M/2$ separated by a distance r.

HINT: We estimated the mass and radius of the Moon in an earlier problem.

HINT: The gravitational potential energy of two masses, M and m, separated by a distance r is $PE_g = -GMm/r$.

ANSWER: In the previous question, we looked at how the gravitational energy of the Sun would change if it shrank dramatically. We approximated this by modeling the Sun as two halves separated by a distance r, and then decreased r dramatically. In order to estimate the energy needed to split an object, we will make the same approximation and instead increase r dramatically. Here we go.

We estimated the mass and radius of the Moon in a previous problem as $M_{Moon} = 6 \times 10^{22}$ kg and $r_{Moon} = 2 \times 10^6$ m. Thus the current gravitational potential energy of the Moon is approximately

$$PE = -\frac{G(M_{\text{Moon}}/2)^2}{r_{\text{Moon}}}$$

$$= -\frac{(7 \times 10^{-11} \text{ N-m}^2/\text{kg}^2)(3 \times 10^{22} \text{ kg})^2}{2 \times 10^6 \text{ m}}$$

$$= -3 \times 10^{28} \text{ J}.$$

In order to split the Moon, we will increase the separation of the two halves from $r = 2 \times 10^6$ m to infinity.* This would increase the potential energy to zero, so we would have to provide $E = +3 \times 10^{28}$ J. This seems like a lot.

Now of course we need to ask "compared with what?" We can first consider it on a human scale. The United States uses about a terawatt (10^{12} W) of electrical power. It would take $t = (3 \times 10^{28} \text{ J})/(10^{12} \text{ W}) = 3 \times 10^{16}$ s $= 10^9$ years to generate that much energy. A megaton of TNT contains 4×10^{15} J, so we would need about 10^{13} large nuclear bombs. Okay. Unsurprisingly, this is a huge amount of energy on a human scale.

---

* If we're impatient, we only need to increase the separation of the two halves by a factor of ten. Getting all the way to infinity can take a very long time [27].

Let's look on a cosmic scale and compare it with the energy of a large meteor. Meteors also orbit the Sun, and so have speeds at Earth orbit that are similar to that of the Earth, or about 3×10^4 m/s. Their kinetic energy is $KE = \left(\frac{1}{2}\right) mv^2$, so the mass of a meteor needed to split the Moon is

$$m_{\text{meteor}} = \frac{PE_g}{(1/2)v^2}$$

$$= \frac{3 \times 10^{28} \text{ J}}{0.5 \times (3 \times 10^4 \text{ m/s})^2}$$

$$= 6 \times 10^{19} \text{ kg}.$$

At a density of 3×10^3 kg/m³, this gives a volume of $V = m/\rho = (6 \times 10^{19} \text{ kg})/(3 \times 10^3 \text{ kg/m}^3) = 2 \times 10^{16} \text{ m}^3 = 2 \times 10^7 \text{ km}^3$. That is a meteor that is between 100 and 1,000 km on a side. That's not just any meteor; that's larger than most of the moons in the solar system.

How large a collision would it take to split
Mars's moon Phobos? In other words, how
much energy would it take to split Phobos into
two approximately equal pieces and separate those
pieces?

ččččččččččččččččččččččč

HINT: Read the previous question first.

HINT: Phobos cannot be seen with the naked eye, but it
can be seen with a reasonable telescope.

HINT: We determined the sensitivity of the naked eye in
an earlier question.

HINT: Phobos is illuminated by the Sun.

HINT: Mars is about 50% further from the Sun than the
Earth is.

ANSWER: In the previous question we estimated the energy needed to split the Earth's Moon. Now let's pick on a moon that's closer to our own size. The next nearest moon to us is Phobos (or Deimos, Mars's other moon). Let's estimate its size and mass. We can estimate the size of Phobos from the fact that it is not visible with the naked eye but can be seen with a reasonably good telescope. Because we have already estimated the sensitivity of our eyes, by estimating how much light would reach our eyes from different size objects at Phobos's distance, we can place an upper limit on the size of Phobos.

Mars (and hence Phobos) is the fourth planet from the Sun. If the inner planets were equally spaced, then Mars would be about 1.3 AU (astronomical unit) from the Sun and between 0.3 and 2.3 AU from the Earth. The planets are, of course, not evenly spaced, and Mars is actually 1.5 AU from the Sun. Even so, that estimate of the minimum distance from the Earth to Mars is off by less than a factor of two.

Given that the solar power density decreases as the square of the distance, Mars gets half as much sunlight as the Earth, or only $p_{Sun} = 500$ W/m^2. The reflected solar power will be the power density times the reflecting area of Phobos. This power will spread out over the area of a sphere of radius $r_{E-M} = 0.5$ AU as it travels 0.5 AU back to our eyes. We also estimated the minimum power density that we can detect as $p_{min} = 2 \times 10^{-12}$ W/m^2 (see "Light a single candle" in chapter 4). This gives

$$p_{min} = p_{Phobos} = \frac{p_{Sun} A_{Phobos}}{4\pi r_{E-M}^2}$$

so that

$$A_{\text{Phobos}} \leq \frac{p_{\min} 4\pi r_{E-M}^2}{p_{\text{Sun}}}$$

$$\leq \frac{(2 \times 10^{-12}\ \text{W/m}^2) \times (4\pi(0.5 \times 1.5 \times 10^{11}\ \text{m})^2)}{500\ \text{W/m}^2}$$

$$\leq 4 \times 10^8\ \text{m}^2$$

$$A_{\text{Phobos}} \leq (2 \times 10^4\ \text{m})^2,$$

or about 20 km in diameter. Note that if we include the reflectivity of Phobos and the effects of atmospheric absorption, then our upper limit will be larger. Let's increase the length limit to 50 km to account for these factors.

Because Phobos can be seen with a reasonable telescope (better than Galileo's and worse than Mt. Palomar's), the lower limit on its reflective area is probably about one hundred times smaller than the upper limit. Reducing the area by a factor of one hundred will reduce the diameter by a factor of ten. This means that its diameter is between 5 and 50 km, so we will take the geometric mean and estimate 20 km. Both Deimos ($d = 8$ km) and Phobos ($d = 25$ km) fall between our upper and lower limits.

Note that, if we were familiar with astronomical visual magnitudes, we could have estimated Phobos's size (relative to Mars) from the apparent magnitude of Mars (about $m = -2$), the limit of human perception ($m = 6.5$), and the fact that luminosity decreases by a factor of ten for each increase of 2.5 in magnitude. If Phobos has luminosity $m = 8$, then it is 10^4 times less bright than Mars with 10^4 times less area. As Mars has a radius of about $r = 4 \times 10^3$ km (less than the Earth's and more than the Moon's), that would give an upper limit for Phobos's radius of 40 km. By the same telescope argument, the lower limit would be 4 km.

Now that we know the diameter, we can estimate the density to determine the mass using $m = \rho V$. The small moons are probably rocky, so they will have a density between that of water ($\rho = 10^3 \text{ kg/m}^3$) and iron ($\rho = 10 \times 10^3 \text{ kg/m}^3$), or about $\rho = 3 \times 10^3 \text{ kg/m}^3$.[*]

$$M_{\text{Phobos}} = \rho V = \rho \frac{4}{3} \pi r^3$$

$$= 4(3 \times 10^3 \text{ kg/m}^3)(10^4 \text{ m})^3$$

$$= 10^{16} \text{ kg}.$$

Now we have all the information we need to estimate the energy needed to split Phobos. The current gravitational potential energy of Phobos is approximately

$$PE = -\frac{G(M_{\text{Phobos}}/2)^2}{r_{\text{Phobos}}}$$

$$= -\frac{(7 \times 10^{-11} \text{ N-m}^2/\text{kg}^2)(5 \times 10^{15} \text{ kg})^2}{10^4 \text{ m}}$$

$$= -2 \times 10^{17} \text{ J}.$$

In order to split Phobos, we would need to increase the separation of the two halves from 10^4 m to infinity. This would increase the potential energy to zero, so we would have to add $+2 \times 10^{17}$ J. This is about ten orders of magnitude less than that needed to split the Moon.

This is a lot of energy, but only on a human scale. It is equivalent to less than one year's electrical energy

---

[*] The latest measurements indicate that the density of Phobos is $1,876 \pm 20 \text{ kg/m}^3$. This stunning precision comes from measuring the gravitational effect of Phobos on spacecraft orbits [28].

production of the United States and only about one hundred 1-megaton bombs. Of course, we would have to get to Phobos first.

Note that, as a practical matter, we would also have to include the effects of Mars's gravity. However, this is a not a very helpful use of the word "practical."

The Sun rotates on its axis once a month. Neutron stars spin around in milliseconds. Ice skaters spin faster and slower by pulling in and extending their arms. Divers and gymnasts spin faster by tucking in their legs. These are examples of the conservation of "angular momentum." The term "conservation," implies that the total amount of angular momentum in the universe does not change. We have explored many of the consequences of energy conservation; now we will explore a few of the consequences of angular momentum conservation.

The angular momentum of an object is the product of its "rotational inertia" and "rotational velocity." Just as inertia (mass) measures the difficulty of changing the velocity of an object (accelerating it), rotational inertia measures the difficulty of changing the rotational velocity (colloquially called "spin") of an object. The greater the mass of an object, the harder it is to change its spin. The further the mass is from the axis of an object, the harder it is to change its spin. For example, it is easy to spin a heavy iron pipe along its long axis (i.e., roll it), when all its mass is close to the axis. Then its mass travels in small circles. It is much more difficult to spin the same pipe perpendicular to its long axis when much of its mass is far from the axis of rotation. Then its mass travels in large circles. The formal expression for rotational inertia (sometimes called "moment of inertia") is

$$I \propto mr^2,$$

where m is the object's mass and r is the typical perpendicular distance from the axis of rotation to the farthest mass. Depending on the shape and mass distribution, there is a proportionality factor that varies from about one-half to one. We will ignore this factor. In addition, because we'll only be dealing with spherical stars and planets, r will always be the radius of the object.

The rotational velocity, ω, is how fast the object spins around a given axis. As that is too simple, instead of measuring it in simple units like rotations per second, physicists make it more complicated by measuring it in radians per second. However, one rotation per second equals 6 radians per second, so it is not that much more complicated.[*]

The angular momentum of a spinning solid object is

$$L = I\omega,$$

where ω is the rotational velocity in radians per second and I is the rotational inertia.

An isolated object cannot change its angular momentum. However, it can change its rotational inertia and its rotational velocity. If its rotational inertia decreases, then its rotational velocity will increase, just as happens when a spinning ice skater pulls in her arms.

Now there is just one more wrinkle. A moving small object, such as a baseball, also has angular momentum. Imagine a baseball bat resting on an icy pond. If a baseball hits it off center, then the baseball bat will spin. The further off center the collision, the faster the bat will spin. Because we know that angular momentum is conserved (that the total angular momentum of baseball plus bat before the collision equals their total angular momentum after the collision), the baseball must have had angular momentum before the collision. This angular momentum is fortunately

[*] To be precise, one rotation per second equals 2π radians per second.

simpler than the other type. The angular momentum of a small object hitting a larger one is

$$L = mvr,$$

where m is the mass of the small object, v is its velocity, and r is the minimum distance between its path and the axis of rotation of the larger object. For example, suppose a meteorite hits the Earth on the equator. If it hits the ground perpendicularly, then it will not change the Earth's rotation. Because its path would intersect the Earth's axis, $r = 0$, and thus its angular momentum with respect to the Earth's axis is zero, $L = 0$. If it hits the ground tangentially, then $r = r_{\text{Earth}} = 6 \times 10^6$ m, and it will add its angular momentum to that of the Earth. Looking down on the Earth from above the North Pole, the two cases would look like this:

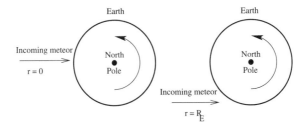

Shrinking Sun

How fast would the Sun
rotate if it collapsed
to a neutron star?
Assume that
the mass stays
the same but
the radius
decreases.

ٿ ٿ

HINT: Angular momentum is conserved.

HINT: The rotational inertia will decrease so that the
rotational velocity must increase by the same factor.

HINT: The Sun rotates on its axis approximately once a
month.

HINT: We estimated the change in the Sun's radius as it
shrinks to a neutron star in a previous question.

HINT: The mass of the Sun does not matter.

ANSWER: This problem is just like the spinning ice skater. Angular momentum, $L = I\omega$, is conserved. When she pulls in her arms, she reduces her rotational inertia, I. This causes her rotational speed, ω, to increase in order to keep her angular momentum unchanged. We can rewrite this as

$$L = I_{old}\omega_{old} = I_{new}\omega_{new}$$

$$\omega_{new} = \omega_{old} \times \frac{I_{old}}{I_{new}} = \omega_{old} \times \frac{M_\odot r_\odot^2}{M_\odot r_{ns}^2}$$

$$= \omega_{old} \times \frac{r_\odot^2}{r_{ns}^2}.$$

This means that we need to estimate the Sun's radius, $r_\odot$; the neutron star's radius, r_{ns}; and the Sun's rotational speed, ω_{old}. Fortunately we have already estimated the radii earlier in this chapter as $r_\odot = 10^9$ m and $r_{ns} = 10^4$ m. The rotational speed will then increase by a factor of $(r_\odot/r_{ns})^2 = 10^{10}$.

Now we just need to estimate the Sun's rotational speed. If we have looked at sunspots, we know that they remain visible on the Sun's surface for many days. This implies that the Sun rotates about once a month (more than once a week and less than once a year). If we haven't ever looked at sunspots, then (a) it is much harder to estimate the Sun's rotational speed and (b) we are missing an interesting experience.

Thus,

$$\omega_{new} = \omega_{old} \times \frac{r_\odot^2}{r_{ns}^2}$$

$$= \frac{1 \text{ rotation}}{1 \text{ month}} \times \frac{(10^9 \text{ m})^2}{(10^4 \text{ m})^2}$$

$$= \frac{10^{10} \text{ rotations}}{30 \times 10^5 \text{ s}}$$

$$= 3 \times 10^3 \text{ rotations/s}.$$

or one rotation in 0.3 ms. Wow. That is fast.

The existence of neutron stars was deduced from the observation of pulsars, objects that emit a powerful radio signal that repeats every few milliseconds.

Spinning Earth

What is the angular momentum of the Earth
as it spins on its axis?

HINT: The Earth completes one rotation a day.

HINT: The mass of the Earth is $M_{Earth} = 6 \times 10^{24}$ kg.

HINT: There are 10^5 seconds in a day.

ANSWER: This is more of a straightforward calculation than an estimation. In order to estimate the angular momentum of the Earth as it spins on its axis, we need to know its mass, radius, and rotational velocity. The rotational velocity is easiest:

$$\omega = \frac{2\pi \text{ radians}}{1 \text{ day}} = \frac{2\pi \text{ radians}}{10^5 \text{ s}} = 6 \times 10^{-5} \text{ rad/s}.$$

Fortunately, we wrote the mass and radius of the Earth on our shirt cuff in order to solve previous questions. After glancing down we know that $M_{Earth} = 6 \times 10^{24} \text{ kg}$ and $r_{Earth} = 6 \times 10^6 \text{ m}$. This gives the rotational inertia of the Earth as

$$I = mr^2 = (6 \times 10^{24} \text{ kg})(6 \times 10^6 \text{ m})^2$$
$$= 2 \times 10^{38} \text{ kg-m}^2.$$

Now we can calculate the Earth's angular momentum as

$$L = I\omega = (2 \times 10^{38} \text{ kg-m}^2)(6 \times 10^{-5} \text{ rad/s})$$
$$= 10^{34} \text{ kg-m}^2/\text{s}.$$

The next questions we should ask are "How big is this?" and "Compared with what?" However, let's wait to make comparisons until we calculate a few more angular momentums.

The dinosaur killer and the day

We think that a large meteorite wiped out the
dinosaurs. By how much could a 1 km meteorite
impact change the length of the day?

HINT: Compare the angular momentum of the meteorite
as it hits the Earth with the angular momentum of the
Earth.

HINT: A meteorite has a density that of between water
and iron.

HINT: All objects in the solar system have about the same
speed when they are 1 AU from the Sun (including the
Earth).

ANSWER: When a meteorite hits the Earth, it adds its angular momentum to that of the Earth. It contributes the maximum angular momentum to the Earth when it hits the Earth tangentially at the equator. Because angular momentum $L = mvr$, we need to estimate the meteorite's mass and speed.

Let's maximize the mass by choosing an iron meteorite (or by having an iron meteorite choose us). Then its mass will be

$$m = \rho V = (10^4 \, \text{kg/m}^3)(1 \, \text{km})^3$$
$$= (10^4 \, \text{kg/m}^3)(10^3 \, \text{m})^3 = 10^{13} \, \text{kg}.$$

Because the meteorite and the Earth both orbit the Sun, when they are at the same distance from the Sun they will have about the same speeds. The Earth's speed around the Sun is

$$v = \frac{2\pi \, 1 \, \text{AU}}{1 \, \text{year}}$$
$$= \frac{2\pi (1.5 \times 10^{11} \, \text{m})}{\pi \times 10^7 \, \text{s}}$$
$$= 3 \times 10^4 \, \text{m/s},$$

so the relative speed of the Earth and the meteorite will be between zero and 6×10^4 m/s, depending on whether it is a following or a head-on collision. Again, let's choose the larger value for the maximum, umm, impact.

The angular momentum of the meteorite is now

$$\mathcal{L}_{\text{meteorite}} = mvr$$
$$= (10^{13} \, \text{kg})(6 \times 10^4 \, \text{m/s})(6 \times 10^6 \, \text{m})$$
$$= 4 \times 10^{24} \, \text{kg-m}^2/\text{s}.$$

Now that we have estimated two angular momentums, we can compare them. The angular momentum of the Earth as it rotates on its axis is $L_{Earth} = 10^{34}$ kg-m^2/s, or about 10^{10} times larger. This means that the meteorite impact would change the Earth's angular momentum, and hence the length of the day, by one part in 10^{10}. As there are 10^5 seconds in a day, this would change the length of the day by about 10^{-5} s or $10\,\mu$s.

Even if it was a 10-km asteroid with 10^3 times more mass, it would only change the length of the day by 0.01 s.

The dinosaurs probably overlooked this particular effect of the collision.

Yellowstone lies on top of a hot spot in the Earth's crust and has produced three supervolcano eruptions in the past few million years. By how much would a supervolcano eruption change the length of the day?

HINT: Compare the angular momentum of the volcano ejecta with the angular momentum of the Earth.

HINT: Mount Saint Helens ejected about 1 km³ of material.

HINT: Estimate the mass that is redistributed to higher or lower latitudes by an eruption.

ANSWER: In the previous question, we looked at the change in the Earth's angular momentum caused by something hitting it. This time, we are looking at the change in the Earth's angular momentum caused by something leaving it. However, we know that a Yellowstone supervolcano eruption would have much less effect than a 10-km meteorite strike because past supervolcano eruptions have had far less effect on the biosphere. This implies that the energy released would be far less.

Also, in order for a supervolcano to change the angular momentum of the Earth, it has to eject material away from the Earth. If the material lands back on Earth, then any angular momentum change caused by the material being launched would be canceled when it lands. There are two reasons why it is unlikely that a volcanic eruption (even a super one) would send significant amounts of material into Earth orbit. The first is that the ejecta would need to reach speeds of 7 km/s or higher. This seems rather improbable. The second is that if this had ever happened, so much debris in Earth orbit would have wrecked our communication and other satellites.

Let's try to estimate the parameters of a supervolcano. Mount Saint Helens ejected about $1 \, km^3$ of material. A supervolcano would eject about 10^3 times as much. This is the same size as a 10-km meteorite. However, the speed of the ejected mass would be much less than the 30 to 60 km/s of the meteorite. Because the material does not achieve Earth orbit, its speed must be less than 7 km/s. The speed is almost certainly larger than the speed of sound, 300 m/s, so we can take the geometric mean and estimate that $v = 2 \, km/s$. Most of the eruption will be upwards, so we'll assume that 10% is directed tangentially. This means that the angular momentum of the ejected material (while it is in flight) is

$$L = mvr = (\rho V)vr$$
$$= 0.1(3 \times 10^3 \text{ kg/m}^3)(10^4 \text{ m})^3(2 \times 10^3 \text{ m/s})$$
$$\times(6 \times 10^6 \text{ m})$$
$$= 4 \times 10^{24} \text{ kg-m}^2/\text{s}.$$

This is 10^{10} times smaller than the angular momentum of the Earth.

There is one other way that the supervolcano could change the length of the day. If it cannot change the angular momentum of the Earth, it can change the mass distribution of the Earth. If the material is ejected from a high latitude and ends up at the equator (or vice versa), then it will change the rotational inertia of the Earth. In order to estimate an upper limit on this effect, let's move all of the mass ejected in the eruption from the equator to the poles. At the equator that mass contributes $I = mr^2$ to the rotational inertia. At the poles that mass is on the axis of rotation so that $r = 0$ and it contributes nothing to the rotational inertia:

$$\Delta I = mr^2 - 0 = m(r_{\text{Earth}})^2$$
$$= (3 \times 10^3 \text{ kg/m}^3)(10^4 \text{ m})^3(6 \times 10^6 \text{ m})^2$$
$$= 10^{29} \text{ kg-m}^2.$$

Even with these extreme assumptions, this decreases the rotational inertia of the Earth by less than one part in 10^9. Thus, it would only increase the rotational velocity of the Earth by less than one part in 10^9. This would decrease the length of the day by less than

$$t = \frac{10^5 \text{ s/day}}{10^9} = 10^{-4} \text{ s/day}.$$

If the Yellowstone supervolcano erupted, we would have more important problems to worry about than a 100-microsecond change in the length of the day.

The orbiting Moon

7.15

What is the angular momentum of the Moon
as it orbits around the Earth?

HINT: The Moon rotates around the Earth once a month.

HINT: The mass of the Moon is $M_{Moon} = 8 \times 10^{22}$ kg.

HINT: The distance from the Earth to the Moon is about
4×10^8 m.

HINT: There are 10^5 seconds in a day.

ANSWER: Because the radius of the Moon is much smaller than the distance from the Earth to the Moon, we can ignore its size. That means that its angular momentum is $L = mvr$, so we need to estimate its mass, speed, and distance from the Earth. We previously estimated the mass of the Moon as $M_{Moon} = 8 \times 10^{22}$ kg.[*] The distance from the Earth to the Moon is $r_{EM} = 4 \times 10^8$ m. The Moon's speed is the distance traveled in one orbit divided by the time it takes, or

$$v = \frac{2\pi r_{EM}}{t},$$

where $t = 1$ month[†] $= 30 \times 10^5$ s $= 3 \times 10^6$ s.

Thus, the angular momentum of the Moon as it orbits the Earth is

$$
\begin{aligned}
L = mvr &= M_{Moon} v r_{EM} \\
&= \frac{M_{Moon} 2\pi r_{EM}^2}{t} \\
&= \frac{(8 \times 10^{22} \text{ kg}) 6 (4 \times 10^8 \text{ m})^2}{3 \times 10^6 \text{ s}} \\
&= 3 \times 10^{34} \text{ kg-m}^2/\text{s}.
\end{aligned}
$$

This is several times larger than the Earth's angular momentum as it spins on its axis.

---

[*] Okay, Okay. We estimated its mass as 6×10^{22} kg. But let's use the correct value this time.

[†] No, we do not care whether the month is a sidereal month, a lunar month, February, or December.

The shortest day

7.16

The average Earth-Moon distance is slowly increasing. This means that the Moon used to be closer to the Earth. We estimated the shortest distance that the Moon could be to the Earth without being torn apart by tides (Roche's Limit). What was the length of the day when the Moon was at Roche's Limit?

HINT: This is the most complicated problem in the book.

HINT: The total angular momentum of the Earth-Moon system is conserved: $L_{Earth} + L_{Moon} = $ constant.

HINT: The closer the Moon is to the Earth, the faster it orbits. The orbital time increases with distance as $T^2 \propto r^3$.

HINT: The closer the Moon is to the Earth, the smaller its angular momentum. Therefore the Earth's angular momentum will be larger.

HINT: The greater the Earth's angular momentum, the faster it rotates on its axis and the shorter the day.

ANSWER: The average distance from the Earth to the Moon increases by about 4 cm each year. The Earth's tidal force would have converted the Moon into a lovely set of rings if the Moon was ever closer to the Earth than Roche's Limit or $r_{min} = 10^7$ m. For purposes of this question we'll assume that the Moon formed at this minimum distance and has been moving away from the Earth ever since.

We want to estimate the length of the day when the Moon orbited the Earth at Roche's Limit. The total angular momentum of the Earth plus the Moon is conserved and does not change. Therefore, we need to know the current angular momentum of the Earth plus the Moon (which we have already estimated) and the angular momentum of the Moon when it was much closer to the Earth.[*]

Currently, we have

$$L_{total} = L_{Moon} + L_{Earth}$$
$$= 3 \times 10^{34} \text{ kg-m}^2/\text{s} + 10^{34} \text{ kg-m}^2/\text{s}$$
$$= 4 \times 10^{34} \text{ kg-m}^2/\text{s},$$

so we need to estimate the angular momentum of the Moon in the distant past.

Note that the Earth currently has one-quarter of the total Earth-Moon angular momentum. At most the Earth could have only four times more angular momentum than today. This would give a "day" that is four times shorter than today.

Today it takes the Moon one month to orbit the Earth. When the Moon was closer to the Earth it took less time.[†] As the angular momentum depends on

---

[*] We are ignoring the angular momentum of the Moon as it spins (rotates) on its axis. Because its mass and radius are both much smaller than the Earth's, its rotational angular momentum must be much, much smaller than the Earth's.

[†] However, it would still have been called a "month."

both distance and time,

$$L = mvr = m(2\pi r / T)r = \frac{2\pi m r^2}{T},$$

we need to determine the exact relationship between the orbital distance r and the time period T. We can determine this in a few ways. We might remember Kepler's Law, that the square of the orbital period is proportional to the cube of the distance, $T^2 \propto r^3$. Alternatively, we might use the fact that the Earth's gravity provides the centripetal force to keep the Moon in its orbit and then solve the physics equations. However, there is a much easier way to approximate the answer.

The minimum Earth-Moon distance of $r_{min} = 10^7$ m is only two times larger than the distance to satellites in low Earth orbit (LEO). This means that we can use the radius ($r = r_E = 6 \times 10^6$ m) and time period ($T = 90$ minutes $= 5 \times 10^3$ s) for these satellites to approximate the Moon's angular momentum at $r = 10^7$ m.[*] Thus,

$$L^{min}_{Moon} = \frac{2\pi m r^2}{T} = \frac{6 \times (8 \times 10^{22} \text{ kg}) \times (6 \times 10^6 \text{ m})^2}{5 \times 10^3 \text{ s}}$$
$$= 4 \times 10^{33} \text{ kg-m}^2/\text{s}.$$

Therefore, if the Moon's orbital distance decreases by a factor of about seventy (from 4×10^8 m to 6×10^6 m), its angular momentum decreases by a factor of about seven. If we decrease the orbital distance by a factor of "only" forty (to 10^7 m), then its angular momentum will still decrease by a factor of five or six.

[*] If you don't remember that satellites in low Earth orbit have a period of 90 minutes, then you will need to go through the longer derivation. I'm sorry.

In this case,

$$L_{\text{Earth}}^{\max} = L_{\text{total}} - L_{\text{Moon}}^{\min}$$

$$= 4 \times 10^{34} \text{ kg-m}^2/\text{s} - \frac{1}{5}(3 \times 10^{34} \text{ kg-m}^2/\text{s})$$

$$= 3 \times 10^{34} \text{ kg-m}^2/\text{s},$$

so the Earth would have three times more angular momentum than today. This means that the day was three times shorter, or about 8 hours.

That was the easy method. Now let's balance the Earth's gravity with the centripetal force to keep the Moon in its orbit and then solve. If you remembered the orbital period of LEO, feel free to skip this section.

$$\frac{G M_{\text{Earth}} M_{\text{Moon}}}{r^2} = \frac{M_{\text{Moon}} v^2}{r},$$

where the Earth-Moon distance is r and the Moon's velocity $v = 2\pi r / T$ so that

$$\frac{G M_{\text{Earth}}}{r^2} = \frac{4\pi^2 r}{T^2}$$

$$T^2 = r^3 \times \frac{40}{G M_{\text{Earth}}}.$$

This is just Kepler's Law, $T^2 \propto r^3$ (for the special case of circular orbits).

The angular momentum of the Moon around the Earth is (from above)

$$L_{\text{Moon}} = mvr = \frac{m 2\pi r^2}{T}.$$

Using the fact that $T \propto r^{3/2}$ and ignoring the factors that do not change, we have

$$L_{\text{Moon}} \propto \frac{r^2}{T}$$

$$\propto \frac{r^2}{r^{3/2}}$$

$$\propto \sqrt{r}.$$

When the Moon orbited the Earth at a distance of only $r_{\min} = 10^7$ m, the distance between them was forty times smaller than it is today. This means that $L_{\text{Moon}} \propto \sqrt{r}$ was $\sqrt{40} = 6$ times smaller, or

$$L_{\text{Moon}}^{\min} = \frac{3 \times 10^{34} \text{ kg-m}^2/\text{s}}{6} = 5 \times 10^{33} \text{ kg-m}^2/\text{s}$$

and

$$L_{\text{Earth}}^{\max} = L_{\text{total}} - L_{\text{Moon}}^{\min}$$

$$= 4 \times 10^{34} \text{ kg-m}^2/\text{s} - 5 \times 10^{33} \text{ kg-m}^2/\text{s}$$

$$= 4 \times 10^{34} \text{ kg-m}^2/\text{s}$$

$$= 4 \times L_{\text{Earth}}^{\text{today}}.$$

Thus, when the Moon was at Roche's Limit, the Earth's angular momentum was four times greater than it is today. Because $L_{\text{Earth}} = I\omega$, the rotational speed of the Earth must have been four times greater, so the Earth rotated four times in 24 hours.

In other words, when the Moon was at Roche's Limit, the day was only 6 hours long. If you worked an 8-hour day, then you had to start work 2 hours before the end of the previous day.

The month was also much shorter. Because $T \propto (r_{EM})^{3/2}$, when the distance was forty times smaller, T was three hundred times smaller. Instead of being 3×10^6 s, the month was only 10^4 s, or about 3 hours.[*]

The tides would have been particularly ferocious. Tidal forces increase as $(r_{EM})^{-3}$, so the tidal force exerted on the Earth by the Moon would have been $(40)^3 = 6 \times 10^4$ times larger. In addition, instead of two tides every 24 hours, there would have been two tides every 3 hours. Those would have really been "tidal" waves.

The surfing must have been amazing!

[*] It is very unlikely that the Moon was ever this close to the Earth. Today the day is shorter than the month. The Earth's rotation pushes the tidal bulge (the extra water at high tide) ahead of the Moon. The bulge gravitationally attracts the Moon and speeds it up. This extra energy moves the Moon slowly outward. If the day were shorter than the month, the effect would be reversed, and the Moon would be slowly pulled inward. But because the day is now shorter than the month, it must always have been shorter. Alas.

Materials

Chapter 8

¿ ¿

All materials are held together by the forces between atoms. By understanding the scale of these forces and the distances between atoms, we can understand the macroscopic forces holding materials together.

The important numbers are the typical (and, I hope, familiar) atomic binding energy of 1.5 eV and the typical atomic size of 10^{-10} m.

Stronger than spider silk

8.1

What is the maximum strength of a material? In other words, what is the maximum weight that a 1-cm cable or rope could possibly support?

ℰ ℰ

HINT: How many atoms are there in the cross section of a 1-cm cable?

HINT: That energy is the force exerted by one atom to another, multiplied by the distance the other atom moves: $E = Fd$.

HINT: The maximum energy released in a chemical reaction is about 1.5 eV.

HINT: An atom is about 10^{-10} m in size.

ANSWER: When we pull on a rope, it stretches. The distance between its atoms increases, and the atoms pull back. The strength of the rope will be related to the force each atom can exert and the number of atoms in the rope.

In order to estimate the maximum strength of a material, we need to estimate how hard each atom can "pull back." This is the maximum force exerted by one atom on its neighbor, the force exerted by a chemical bond. We already estimated the maximum energy of a chemical bond to be about 1.5 electron volts (1.5 eV) when we estimated the value of recycling an aluminum can. We just need to relate energy to force. Fortunately, we already looked at this back in chapter 5. If a force F is applied over a distance d, then it transfers energy (does work) of

$$E = Fd.$$

The typical radius of an atom is 10^{-10} m.[*] If the interatomic force is exerted over that distance, then we have

$$
\begin{aligned}
F_{\text{atom}} &= \frac{E}{d} = \frac{1.5\,\text{eV}}{10^{-10}\,\text{m}} \\
&= \frac{(1.5\,\text{eV})(1.6 \times 10^{-19}\,\text{J/eV})}{10^{-10}\,\text{m}} \\
&= 2 \times 10^{-9}\,\text{J/m} = 2 \times 10^{-9}\,\text{N}.
\end{aligned}
$$

That is much smaller than the 10^{-1} N we can detect. However, compared with the mass of a typical atom, $m = 10^{-26}$ kg, it is enormous. Of course, both of those comparisons are somewhat irrelevant.

Now that we know the force exerted by one atom, we need to estimate the number of atoms in the rope. Because the length of an ideal rope does not affect

---

[*] In fact, all atoms are about the same size. Even though their number of electrons can vary by a factor of one hundred, their radius only varies by a factor of a few.

its strength, we just need to estimate the number of atoms in a cross section of the rope. We simply need to cut the rope at some point and count the number of atoms. If we can count one atom per second, we should be able to finish this before the Sun burns out.

Okay, let's try a different method. Let's choose a rope that is 1 m thick, with a square cross section. Although this is a rather thick rope with nasty corners, it is a convenient size for calculations. We can try more manageable ropes later. Given that each atom has a radius of $r = 10^{-10}$ m, there are

$$n = \frac{1\,\text{m}}{2 \times 10^{-10}\,\text{m}} = 5 \times 10^9$$

atoms along one side of the rope. This gives a total of

$$N = (5 \times 10^9)^2 = 3 \times 10^{19}$$

atoms in a square-meter of rope.

The total force exerted by all those atoms will be

$$\begin{aligned}F_{\text{tot}} &= N F_{\text{atom}} \\ &= (3 \times 10^{19}\,\text{atoms})(2 \times 10^{-9}\,\text{N/atom}) \\ &= 6 \times 10^{10}\,\text{N}.\end{aligned}$$

The tensile strength would then be

$$T = 6 \times 10^{10}\,\text{N/m}^2.$$

This is the maximum tensile strength of any possible material. It assumes that all of the atoms are pulling their hardest and that there are no defects or other weaknesses in the material. For comparison, carbon nanotube fiber has a maximum strength of about 10^{10} N/m^2 (as of 2007 [29]) and steel piano wire has a maximum strength of about 2×10^9 N/m^2.[*]

---

[*] Note that the superstrong carbon fibers were only 1 mm long. Longer fibers were much weaker. There is still a lot of work to be done to make the ultimate rope.

Now let's consider the maximum load we could hang from a 1-cm rope. The cross-sectional area of the rope is $A = (1\,\text{cm})^2 = 10^{-4}\,\text{m}^2$, so the maximum weight it could support would be

$$F = TA = (6 \times 10^{10}\,\text{N/m}^2)(10^{-4}\,\text{m}^2) = 6 \times 10^6\,\text{N},$$

or the weight of 6×10^5 kg or 600 tons. Imagine supporting 600 tons from a 1-cm rope. For comparison, 1-cm (0.375-inch) of polyethylene or polypropylene rope has a safe working load of only about 100 pounds, or about 10^4 times less.[*]

Now that is strong!

[*] Yes, we are confusing maximum load with safe working load.
However, that is only a factor of three or so difference.

8.2

Beanstalk to orbit

One proposed method to lift people
and cargo into orbit is a beanstalk,
a very long cable stretching from
the Earth's surface out past
geosynchronous orbit and orbiting
the Earth once a day so it stays
tethered over the same point. The
weight of the cable would be
counterbalanced by a large mass
located beyond geosynchronous
orbit. How strong would the
beanstalk cable have
to be to support
its own weight?
The answer
should be
in N/m².

HINT: We can model the cable as just hanging from
geosynchronous orbit.

HINT: Geosynchronous orbit is about 4×10^7 m from the
center of the Earth.

HINT: Carbon nanotubes will have a density of around
1 ton/m³.

HINT: The force of Earth's gravity decreases as $F \propto 1/r^2$,
where r is the distance to the center of the Earth.

ANSWER: Science fiction stories have long proposed lifting people and cargo into orbit using a beanstalk, a very long cable stretching from the Earth's surface out past the height of geosynchronous orbit and orbiting the Earth once a day so it stays tethered over the same point. The parts of the cable closer to the Earth would try to orbit more quickly. They would be counterbalanced by the parts that are further from the Earth, which would try to orbit more slowly. Vehicles could climb up the cable to the height of geosynchronous orbit and then release their payloads into orbit. This would replace all of the extremely inefficient and rather terrifying rockets currently used to reach orbit.[*]

In effect, the beanstalk would "hang" from geosynchronous orbit (GEO). The part closer to the Earth would hang "down" and the part farther from the Earth would hang "up."[†] In order to estimate the strength needed for the beanstalk cable, we need to estimate the height of GEO, the mass of the cable, and the weight of the cable.

Let's start with GEO. We might remember that GEO is 2.5×10^4 miles or 4×10^4 km from the center of the Earth. If not, then we need to do some orbital calculations. An object orbiting the Earth at GEO stays over the same point on the ground, so it takes $T = 1$ day $= 10^5$ s for each orbit. This gives it a speed $v = 2\pi r / T$. The Earth's gravity provides the centripetal force needed to stay in orbit, so

$$\frac{G M_{\text{Earth}} m}{r_{\text{GEO}}^2} = \frac{m v^2}{r_{\text{GEO}}}$$

$$G M_{\text{Earth}} = \left(\frac{2\pi r}{T}\right)^2 r$$

---

[*] The payload of a modern rocket is only about 10^{-2} to 10^{-3} of its mass. The rest is fuel and engine.

[†] So don't try telephoning.

$$r_{GEO} = \left(\frac{T^2 G \, M_{Earth}}{4\pi^2} \right)^{1/3}$$

$$= \left(\frac{(10^5 \text{ s})^2 (7 \times 10^{-11} \text{ N-m}^2/\text{kg}^2)}{40} (6 \times 10^{24} \text{ kg}) \right)^{1/3}$$

$$= (10^{23} \text{ m}^3)^{1/3}$$

$$= 10^8 \text{ m} = 10^5 \text{ km}.$$

This calculation would give the correct answer of 4×10^4 km if we had used the exact length of the day and done the arithmetic more precisely. On the other hand, if we don't remember the distance to GEO and can't be bothered to look it up, that's definitely good enough.

This means that we will need a cable that hangs down for a length of

$$L = r_{GEO} - r_{Earth} = 4 \times 10^4 \text{ km} - 6 \times 10^3 \text{ km}$$

$$= 3 \times 10^4 \text{ km}.$$

That is a rather long cable. If it is 1-m thick, then it will have a volume of

$$V = (1 \text{ m})^2 (3 \times 10^7 \text{ m}) = 3 \times 10^7 \text{ m}^3.$$

Let's make the cable from relatively light carbon nanotubes (rather than much heavier steel), so it will have a mass of

$$m = \rho V = (10^3 \text{ kg/m}^3)(3 \times 10^7 \text{ m}^3) = 3 \times 10^{10} \text{ kg}.$$

If it was all subject to Earth's full (i.e., surface) gravity, then it would have a weight of

$$W = mg = (3 \times 10^{10} \text{ kg})(10 \text{ N/kg}) = 3 \times 10^{11} \text{ N},$$

giving a tensile stress (force per area) of 3×10^{11} N/m^2.

Uh oh. That is five times larger than the maximum possible tensile strength of $T = 6 \times 10^{10}$ N/m² that we estimated in the preceding problem.

Fortunately, the cable will not weigh quite that much, as the Earth's gravitational force drops rapidly with distance. At the surface (a distance of 6×10^6 m from the center of the Earth), the Earth's gravitational field is $g = 10$ N/kg. At GEO, the distance is seven times larger, so the gravitational field is $7^2 = 50$ times smaller or $g = 0.2$ N/kg. Taking the geometric mean, we get an average gravitational field of

$$g_{avg} = \frac{10 \, \text{N/kg}}{7} = 1.5 \, \text{N/kg}.$$

This reduces the total weight of the cable by a factor of seven to

$$W = mg_{avg} = (3 \times 10^{10} \, \text{kg})(1.5 \, \text{N/kg}) = 5 \times 10^{10} \, \text{N},$$

giving a tensile stress (force per area) of 5×10^{10} N/m², which is very close to the maximum material tensile strength.

Now all we need to do is to increase the strength of really strong carbon nanofibers by a factor of six and increase their length from 10^{-3} m to 4×10^7 m. Hey, that's only an increase of 10^{10} in length!

Note that this is one of the few problems in this book that falls in the "Goldilocks" category of "just right." It will take far more effort and analysis to determine whether beanstalks might actually be possible.

Bolt failure

How easy is it to shear a $\frac{5}{16}$-inch screw?
I was repairing my dock and using an 8-inch
wrench to tighten a very stubborn $\frac{5}{16}$-inch lag
screw. I sheared the head of the screw off.
Was I unexpectedly strong, was the screw
unexpectedly weak,
or was this to be
expected?

HINT: The tensile strength of steel is 4×10^8 N/m^2.

HINT: How much force can you apply to a wrench? 1 lb =
5 N.

HINT: What is the cross-sectional area of the screw?

HINT: The force is effectively multiplied by the moment
arm, the distance over which it is applied. What is the
ratio of the moment arms of the force applied by the
wrench and by the screw?

ANSWER: We apply torque to the head of the screw by applying a force at or near the end of the wrench. This torque is opposed by friction between the screw threads and the material it is screwed into. The torque is transmitted along the screw by the body of the screw. Let's model the screw at the location where it sheared as two disks in contact with each other on a common axis. The two disks are held together by the strength of the molecular bonds between them.

Torque equals force times the distance (moment or lever arm) over which it is applied. The longer the wrench, the greater the torque we can apply.

In this case, we are using an 8-inch wrench, so the moment arm for our force is 8 inches (20 cm). The moment arm for the force applied by the screw (i.e., by one disk on the other) is the screw radius. The screw diameter at the shank is $\frac{5}{16}$ inch, so the radius is $\frac{5}{32}$ inch $= 125/32$ mm $= 4$ mm. We can apply a force of about 100 pounds. Because the ratio of moment arms is $(20\,\text{cm})/(4\,\text{mm}) = 50$, the effective force between the two disks is about 50×100 pounds $= 5,000$ pounds, or 3×10^4 N.

This force is applied over an area of

$$A = \pi r^2 = 4(4\,\text{mm})^2 = 60\,\text{mm}^2 = 6 \times 10^{-5}\,\text{m}^2.$$

(Note that we want the area weighted by the moment arm [the distance from the axis], not the area times the moment arm. This will decrease the effective area by a factor of a few). Thus the stress (force per area) is

$$S = \frac{3 \times 10^4\,\text{N}}{6 \times 10^{-5}\,\text{m}^2} = 5 \times 10^8\,\text{N/m}^2.$$

While this is one hundred times smaller than the maximum material strength we estimated in an earlier question, it is only two times smaller than the strength of steel piano wire.

According to the manufacturer, the tensile strength of lag screw steel is 60,000 psi (4×10^3 atm = $4 \times 10^8 \text{N/m}^2$). Thus, I applied a force greater than the screw strength. It is not surprising that the screw sheared.

For the next screw, I drilled a pilot hole first.

Making mountains out of molecules

The Himalayas are rising as India collides with Asia.
How much taller can they get? What is the
maximum height of a mountain on the Earth?
On Mars?

ꞔ ꞔ

HINT: The energy of a typical molecular bond is 1.5 eV.

HINT: The energy required to melt rock is about one-tenth
of the energy of a typical molecular bond.

HINT: Consider a cylindrical mountain.

HINT: When the mountain is too tall, it sinks a little,
transferring energy from gravitational potential energy to
making the rock at the bottom flow.

ANSWER: The tallest mountain on the Earth is Mt. Everest, at an elevation of about 3×10^4 ft, or 10^4 m. It is part of the Himalayas, which are rising as India collides with Asia. Can Mt. Everest get much taller? What is the limit for terrestrial mountains?

When the weight of the mountain becomes too great to be supported by the Earth's crust, the bottom layer of rock will flow, and the mountain will sink. This process is governed by energy. Lowering the mountain releases gravitational energy. The taller the mountain, the greater the gravitational energy released. Making rock flow (i.e., turning it from solid to liquid) requires energy. The mountain is at its maximum height when the gravitational energy released by lowering the mountain is barely less than the energy needed to make the rock flow.

This means that we need to estimate the gravitational energy released and the energy needed to make the rock flow. Rock is generally made of silicon dioxide molecules, SiO_2, with a molecular weight of about 60 g/mole = 0.06 kg/mole. Thus the mass of one SiO_2 molecule is $m = W/N_A = (0.06 \text{ kg/mole})/(6 \times 10^{23}$ molecules/mole$) = 1 \times 10^{-25}$ kg.

The silicon and oxygen in each SiO_2 molecule is tightly bound, presumably with the typical molecular binding energy of 1.5 eV. However, the energy needed to rearrange solid SiO_2 should be much less. We will estimate that it is about ten times less, or 0.15 eV/molecule. For a more detailed discussion of the physics involved, see [30].

Now let's consider a one-molecule-wide column of rock that is the height of the mountain, h. If the entire column sinks by one molecule, that is equivalent to lowering one molecule a distance h from the top to the bottom of the column. The gravitational energy released by that is

$$PE = mgh,$$

where m is the mass of one SiO_2 molecule. In order to do this, one molecule of SiO_2 must move out of the way (flow) at an energy cost of 0.15 eV. The mountain is at its highest when

$$mgh = 0.15\,\text{eV}$$
$$h = \frac{0.15\,\text{eV}}{mg} = \frac{(0.15\,\text{eV})(1.6 \times 10^{-19}\,\text{J/eV})}{(10\,\text{N/kg})(10^{-25}\,\text{kg})}$$
$$= 2 \times 10^4\,\text{m}.$$

This is about twice the current height of Mt. Everest. Given the uncertainty of our result, all we can really say is that Mt. Everest will not get ten times higher.

As this result is inversely proportional to the strength of gravity at the surface, mountains on the Moon can be six times taller and mountains on Mars can be a few times taller. This is consistent with the height of Olympus Mons, which is 22 km tall.

Chopping down a tree

How many bullets does it
take to cut down a tree?

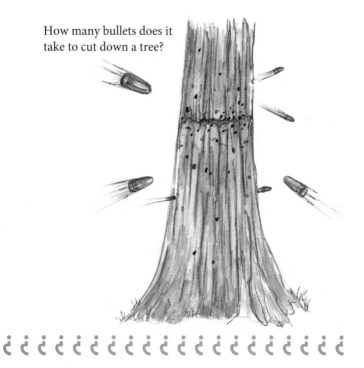

HINT: The bullets have kinetic energy.

HINT: It takes energy to break the chemical bonds that
hold the atoms of the tree together.

ANSWER: When a bullet breaks things, it converts its kinetic energy into chemical energy by breaking the bonds that hold the atoms together. When we chop down a tree, by whatever means (e.g., using axes, bullets, chainsaws, or herrings), we concentrate our efforts on a thin slice of the tree. In order to figure out how many bullets it will take to cut down a tree, we need to estimate the kinetic energy of each bullet as well as the chemical energy of that slice of tree.

Let's choose a tree larger than a sapling and smaller than a redwood, with a radius of 20 cm (8 inches or one hand-span). Why choose 20 cm? Because it is a nice round number and because that is a substantial tree. Let's also assume really good marksmanship so that the bullets will remove a thin slice of tree that is only about 2 cm tall (about 1 inch) for a total removed volume of

$$V = \pi r^2 h = 3(20\,\text{cm})^2(2\,\text{cm}) = 2 \times 10^3\,\text{cm}^3.$$

At a density of $\rho = 1$ g/cm^3, that gives a mass of $m = \rho V = 2$ kg.

The energy needed to break the atomic bonds in that 2-kg slice is about the same as the energy gained by rearranging the atomic bonds in 2-kg of gasoline. At 4×10^7 J/kg (see chapter 3 for more details), this gives a total energy needed to break the bonds of $E = 8 \times 10^7$ J.

Now we need to estimate the kinetic energy of a bullet. A typical lead bullet is around 1 cm in diameter and 2 cm long.[*] This gives a mass of

$$m_b = \rho V = (10\,\text{g/cm}^3)(2\,\text{cm}^3) = 20\,\text{g} = 2 \times 10^{-2}\,\text{kg}.$$

[*] Modern bullets range in size from .22-caliber/5.56 mm to .50-caliber/12.7 mm in diameter. They range in mass from 4 to 50 g.

At a muzzle velocity of 10^3 m/s, the kinetic energy per bullet is

$$KE_b = \frac{1}{2}m_b v^2 = 0.5(2 \times 10^{-2}\,\text{kg})(10^3\,\text{m/s})^2 = 10^4\,\text{J}.$$

In order to cut down a tree, we would then need

$$N = \frac{E}{KE_b} = \frac{8 \times 10^7\,\text{J}}{10^4\,\text{J}} = 10^4\,\text{bullets}.$$

This assumes that we need to break every single atomic bond in that slice of tree and that the bullets are 100% efficient at using their kinetic energy to break bonds. Fortunately, these rather unrealistic assumptions tend to cancel each other.

The Mythbusters chopped down a somewhat smaller pine tree using a minigun that fired 0.30-caliber/7.62 mm bullets at a rate of fifty per second. It took them 45 seconds or about 2×10^3 bullets [31].

Radiation

Chapter 9

Radiation is all around us, yet it is mysterious and scary to many people. As one of my students said, "If it's natural, how can it be bad for you?" In this chapter we will estimate some common, uncommon, and extremely uncommon radioactive phenomena.

First, we need to introduce some concepts. "Radiation" consists of high-speed, high-energy particles. They can cause damage when they interact in material (such as our bodies). There are many sources of radiation, both natural and man-made. "Radioactive materials" emit radiation. Nuclear reactors emit radiation. The Sun, supernovas, and other cosmic sources emit radiation that are called cosmic rays when they hit the Earth. About one cosmic ray passes through your hand every second.

These high-energy particles can be charged or neutral. When charged particles pass through matter, they deposit energy by jiggling[*] the atomic electrons. For every centimeter of water they pass through, they deposit 2 MeV (2×10^6 electron volts) of energy (or 20 MeV per cm of iron or lead). When they run out of energy, they stop.

When neutral particles pass through matter, they only deposit energy when they interact with the atomic nuclei. The probability of this interaction is called the cross section and is measured in units of area.[†] In order to find the probability of interaction in a certain amount of matter, we multiply the cross section times the density of protons and neutrons. For example, the reactor neutrino cross section[‡] was first measured in

[*] That's a technical term.

[†] A very large unit of cross section is the barn, which is equal to 10^{-24} cm^2. I like to think that it refers to being able to hit the broad side of a barn.

[‡] This only applies to neutrinos with energy from radioactive decay with energy from keV to MeV. Cross sections for higher energy neutrinos are much larger.

1956 [32] to be

$$\sigma = 6 \times 10^{-44} \, \text{cm}^2 = 6 \times 10^{-48} \, \text{m}^2.$$

Because the mass of one proton or neutron is $m = (1 \, \text{g/mole})/(6 \times 10^{23} \, \text{mole}^{-1}) = 2 \times 10^{-27} \, \text{kg}$, the number density of protons and neutrons in water is

$$n = \frac{\rho}{m} = \frac{10^3 \, \text{kg/m}^3}{2 \times 10^{-27} \, \text{kg}} = 6 \times 10^{29}/\text{m}^3.$$

This means that the typical distance that a neutrino can travel (its mean free path) through water before interacting is

$$d = \frac{1}{n\sigma} = \frac{1}{(6 \times 10^{29} \, \text{m}^{-3})(6 \times 10^{-48} \, \text{m}^2)}$$

$$= \frac{1}{(4 \times 10^{-18} \, \text{m}^{-1})}$$

$$= 3 \times 10^{17} \, \text{m}.$$

This is a rather long distance. Traveling at the speed of light, $c = 3 \times 10^8$ m/s, it would take 10^9 s, or 30 years (at $\pi \times 10^7$ s/yr), to travel that far. In other words, the interaction distance of neutrinos in water is 30 light-years. The interaction distance in lead, as it is ten times denser, is a mere 3 light-years.

No wonder it took physicists twenty years to detect the neutrino after it was predicted.

Neutrinos are emitted in fusion reactions by the Sun and in fission reactions by nuclear reactors. For every four hydrogen nuclei that the Sun fuses into one helium nucleus, the Sun gains about 30 MeV of energy and emits two neutrinos (as two of the protons are converted to neutrons). For every uranium-235 nucleus that fissions (splits into two smaller nuclei), the reactor gains about 200 MeV of energy, and the daughter nuclei release several neutrinos (and other particles) as they decay further.

Radiation damages our bodies by breaking chemical bonds. We measure the radiation dose in "grays," where

$$1 \text{ Gy} = 1 \text{ J/kg}$$

measures the energy deposited in our bodies by the radiation.* Neutrinos deposit about 1 MeV when they interact. Very high-energy ($E \gg 1$ MeV) charged particles deposit 2 MeV for every cm of tissue they pass through. Acute (as opposed to chronic) radiation poisoning starts at a single dose of about 0.7 Gy. A dose of 10 Gy is almost always lethal.† Radiation doses are typically reported in sieverts rather than grays to account for the different toxicity posed by different types of radiation.

---

* An older unit is the rad where 1 rad = 0.01 J/kg of deposited energy.

† This refers to whole body irradiation by penetrating radiation, with the entire dose acquired in a short period of time. For more information, see the Centers for Disease Control [33].

Nuclear neutrinos

9.1

How many neutrinos from our local nuclear power plant pass through our bodies?

HINT: A big nuclear power plant consumes about 3 GW of thermal power to produce about 1 GW of electrical power.

HINT: Each nuclear fission produces 200 MeV of energy and several neutrinos.

HINT: If we live 100 km from the power plant, those neutrinos will be evenly distributed over a sphere of radius 100 km.

HINT: Our cross-sectional area, as we face the nuclear plant, is about 1 m².

315

ANSWER: Nuclear power plants produce lots of energy by fissioning lots of uranium nuclei. As those nuclei fission, they each release about 200 MeV of energy and several neutrinos.[*] A typical big, commercial nuclear power plant produces about 1 GW of electrical power by consuming about 3 GW of thermal power. In order to produce 3 GW, the power plant has to fission a lot of uranium:

$$N = \frac{(3 \times 10^9 \text{ W})(6 \times 10^{18} \text{ eV/J})}{(2 \times 10^8 \text{ eV/uranium})}$$

$$= 2 \times 10^{20} \text{ uranium/s},$$

where we convert from eV to joules and use the fact that a watt is a joule per second. That is a lot of uranium nuclei, but it is 3×10^3 times smaller than a mole.[†] At several neutrinos per fission, there are about 10^{21} neutrinos released every second.

Now we need to estimate the distance from us to the nearest nuclear power plant. There are about one hundred nuclear power plants in the United States, so it is very likely that we are less than 10^3 km and more than 10 km from the nearest one, giving a geometric mean of 100 km. If your distance is different, feel free to use it.

Then the 10^{21} neutrinos per second will be evenly distributed over the surface of a 100-km sphere with area

$$A = 4\pi r^2 = 12 \times (10^5 \text{ m})^2 = 10^{11} \text{ m}^2,$$

---

[*] The exact number of neutrinos will depend on which daughter nuclei the uranium splits into and how those daughter nuclei decay.

[†] That means that in one year, or 3×10^7 s, a power plant will consume 10^4 moles of uranium. At 0.2 kg/mole, that is 2 tons. That is not much material to power an entire nuclear plant for a year.

so the neutrino flux will be

$$f_\nu = \frac{10^{21} \nu/s}{10^{11} \, m^2} = 10^{10} \nu/s \cdot m^2.$$

Our surface area will depend on our orientation. If we lie down with our feet pointing at the plant then we will minimize our surface area but maximize the distance that each neutrino passes through us. Let's instead face the power plant squarely. In that case our frontal area is about 2 m by 0.5 m, giving an area of $A = 1 \, m^2$.

This means that every second, 10^{10} neutrinos pass through each of our bodies. If we live ten times closer, then there are 10^{12} neutrinos per second, and if we live ten times farther away, then there are a mere 10^8 neutrinos per second. No matter what, that's a lot of neutrinos.

Fortunately, they do not interact much.

How many of the neutrinos from our local nuclear power plant interact in our bodies?

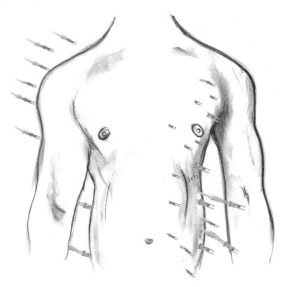

¿ ¿

HINT: What is our average density?

HINT: How does that compare to 30 light-years of water?

HINT: What is our average thickness?

HINT: We estimated the neutrino flux in the previous question as $10^{10}\,\nu$ per second per square meter.

ANSWER: About 10^{10} neutrinos pass through our bodies every second. However, as we showed earlier, a neutrino can travel through about 30 light-years of water before interacting. This means that we need to estimate our thickness, or at least our physical thickness.[*]

Assuming that we are facing the power plant squarely, then our area is about $1\,\text{m}^2$. We can estimate our thickness in a few ways. It is definitely more than 1 cm and less than 100 cm, giving a geometric mean of 10 cm. Or we can measure our thickness in a few places and take the average. However, there is a more reliable method. We have about the same density as water. Assuming for ease of calculation that our mass is 100 kg, this means that our volume is

$$V = \frac{m}{\rho} = \frac{100\,\text{kg}}{10^3\,\text{kg/m}^3} = 0.1\,\text{m}^3.$$

This will be equal to our area times our average thickness, $V = At$. If our area is $1\,\text{m}^2$, then our average thickness is 0.1 m. Note that if we turn sideways so that our area exposed to the neutrinos halves to $0.5\,\text{m}^2$, then our thickness doubles to 0.2 m.

The probability that any one neutrino interacts in our body is the ratio of our thickness and the neutrino interaction length:

$$P = \frac{t}{d} = \frac{0.1\,\text{m}}{3 \times 10^{17}\,\text{m}} = 3 \times 10^{-19}.$$

Note that if we orient ourselves feet first to the power plant, then fewer neutrinos will pass through our bodies, but they will each pass through a greater thickness. The number of neutrinos passing through is proportional to our area, and the interaction probability is proportional to our thickness, so the total number interacting is proportional to our volume (area times

[*] My mental thickness cannot be measured in meters.

thickness). This is why modern neutrino detectors are measured in kilotons rather than in area and thickness.

The total number of neutrinos passing through our bodies in our lifetimes is

$$N = (10^{10} \nu/s)(\pi \times 10^7 \text{ s/yr})(10^2 \text{ yr}) = 3 \times 10^{19}.$$

This means that the total number of neutrinos inter-acting in our bodies in our lifetimes is

$$I = NP = (3 \times 10^{-19})(3 \times 10^{19}) = 10,$$

or about one every ten years.

This must be why those decadal birthdays (30, 40, 50, . . .) are so painful.

How many neutrinos from the Sun pass
through our bodies every second?
How many interact?

¿ ¿

HINT: The solar energy flux is about 10^3 W/m²-s at Earth orbit.

HINT: For every four hydrogen nuclei converted into helium, the Sun releases two neutrinos and 30 MeV of energy.

ANSWER: The Sun produces a lot of energy and a lot of neutrinos. It is also very far away. In order to estimate the number of solar neutrinos passing through our bodies we need to estimate both the total neutrino output and the area over which they are distributed.

Let's start with the neutrino output. As the Sun fuses four hydrogen nuclei into one helium nucleus, it transforms about 30 MeV from mass energy (the binding energy of ^{4}He) into thermal energy, and it releases two neutrinos.* This means that we need to know the power output of the Sun. Astronomers (or readers of *Guesstimation*) might remember that the power output of the Sun is 4×10^{26} W. Astrophysicists might be able to calculate it from the solar surface area and temperature (about 5×10^3 K). We'll just use the solar constant at Earth orbit of 10^3 W/m^2 = 10^3 J/s $\cdot$ m^2. This means that the neutrino flux at Earth orbit is

$$f_v = \frac{\text{solar power}}{\text{area}} \times \frac{\text{neutrino}}{\text{energy}}$$

$$= (10^3 \text{ J/s·m}^2) \times (6 \times 10^{18} \text{ eV/J}) \times \frac{2 \text{ neutrino}}{30 \times 10^6 \text{ eV}}$$

$$= 4 \times 10^{14} \text{ neutrino/s·m}^2.$$

Thus, when we are lying down and sunbathing at high noon or standing up and facing the setting Sun, there are 4×10^{14} neutrinos per second passing through our bodies. This is 4×10^4 more than from that local nuclear power plant.

The probability of a single neutrino interacting in our bodies will be about the same for solar neutrinos

---

* The first step in the reaction chain is to fuse two hydrogen nuclei (protons) into a heavy hydrogen nucleus $p + p \rightarrow {}^2\text{H} + e^+ + v$, where the heavy hydrogen (deuterium) nucleus contains a proton and a neutron.

as for power plant neutrinos. Because 4×10^4 times more solar neutrinos than nuclear plant neutrinos pass through our bodies, there will be 4×10^4 times more solar neutrino interactions. Instead of one every ten years, there will 4×10^3 per year or about ten per day.

Ouch!

HINT: The supernova is powered by the energy released by the gravitational collapse of the star.

HINT: We estimated that energy release in the "Collapsing Sun" question.

HINT: The neutrinos arrive first.

HINT: 99% of the energy is released in the form of neutrinos.

HINT: An exposure of 10 J/kg of energy deposited in our bodies by radiation is lethal.

HINT: Each neutrino has about 1 MeV of energy.

? ?

If the Sun went supernova, what would kill us (first)?

Supernovas can be dangerous

9.4

ANSWER: It is generally not a good idea to stand too close to a supernova. A literally cosmic amount of energy is released in the form of outgoing particles, including neutrinos, photons (light), and regular matter. In a terrestrial explosion, the blast of photons is called the "flash," and the blast of regular matter is called the shock wave.* For some reason we do not have a common expression referring to the blast of neutrinos. However, unlike terrestrial explosions, supernovas release 99% of their energy in the form of neutrinos.

First let's consider the speeds and arrival times of the different particles. The regular matter is clearly the slowest, as it travels slower than the speed of light. The photons and neutrinos both travel at about the speed of light.† The neutrinos travel directly away from their production point until they interact somewhere else. However, the star is opaque, so the photons are repeatedly absorbed and reemitted by the stellar material until they finally reach the surface. This means that the neutrinos arrive first. In fact, for supernova SN 1987A, a pulse of about twenty neutrinos was detected about 3 hours before the first photons arrived (i.e., before the star got measurably brighter).

Given that the neutrinos arrive first, let's estimate that radiation dose. In order to do this, we need to estimate the number of neutrinos emitted, the number passing through our bodies, the number interacting in our bodies, and the energy deposited by those interacting neutrinos. Let's get started.

As we estimated back in the "Collapsing Sun" question, the energy released by gravitational collapse of

---

* On the Earth, the shock wave is actually an atmospheric pressure wave.

† Neutrinos do have a tiny mass and so travel a tiny bit slower than the speed of light. However, that difference is negligible, even over distances of light-years.

our Sun would be 10^{46} J. A star that could supernova would be several times larger than our Sun and so would release several times more energy.[*] The vast majority of that energy will be carried away by the neutrinos. These neutrinos come from the radioactive decay of unstable elements created during the stellar collapse and hence will have about the same energies (about 1 MeV) as the neutrinos from power plants (at least within a factor of ten). Thus, we can calculate the number of neutrinos emitted as

$$N_\nu = \frac{\text{energy released}}{\text{energy per neutrino}}$$

$$= \frac{(10^{46} \text{ J})(6 \times 10^{18} \text{ eV/J})}{10^6 \text{ eV}/\nu}$$

$$= 6 \times 10^{58} \nu,$$

which is a really, really big number compared with pretty much anything.

The number of neutrinos passing through our bodies on the Earth at a distance of $r = 1$ AU $= 1.5 \times 10^{11}$ m from the supernova will be proportional to the area of our bodies divided by the area of a sphere of radius $r = 1$ AU:

$$N_\nu^{\text{body}} = N_\nu \times \frac{\text{body area}}{\text{1 AU sphere surface area}}$$

$$= 6 \times 10^{58} \nu \times \frac{1 \text{ m}^2}{4\pi (1.5 \times 10^{11} \text{ m})^2}$$

$$= 2 \times 10^{35} \nu$$

will pass through our bodies as we watch the supernova.

We already estimated the probability of a single neutrino interacting in our body in an earlier question

[*] No. Our Sun is not large enough to supernova. Now we can sleep quietly tonight.

to be

$$P = \frac{t}{d} = \frac{0.1\,\text{m}}{3 \times 10^{17}\,\text{m}} = 3 \times 10^{-18},$$

where t is our thickness and d is the neutrino interaction length.

Thus, the total number of neutrinos interacting in our bodies will be

$$N_\nu^{\text{interact}} = N_\nu P = (2 \times 10^{35})(3 \times 10^{-18})$$

$$= 6 \times 10^{17},$$

and each interaction will transfer up to 1 MeV of energy. Let's estimate that the average interaction transfers one-tenth of the maximum, or $E_{interact} = 0.1$ MeV. Then the total energy deposited in our bodies by the neutrinos is

$$E = N_\nu^{\text{interact}} E_{\text{interact}} = (6 \times 10^{17})(0.1\,\text{MeV})$$

$$= (6 \times 10^{22}\,\text{eV})(1.6 \times 10^{-19}\,\text{J/eV})$$

$$= 10^4\,\text{J}.$$

Assuming we have a mass of 100 kg, this corresponds to an energy deposited per kilogram (and hence radiation dose) of

$$e = \frac{10^4\,\text{J}}{100\,\text{kg}} = 10^2\,\text{J/kg} = 10^2\,\text{Gy},$$

which is ten times the lethal dose. Death by neutrino!

Supernovas can be beautiful, but don't get too close.

A biologist recently claimed to have revived a 30 million-year-old bacterium. How many cosmic rays would have passed through that bacterium during that time? What is the probability that its DNA was scrambled?

ċ ċ

HINT: One cosmic ray passes through our hand every second.

HINT: Bacteria are about 10^{-6} m in size.

HINT: A cosmic ray is a high-energy, high-velocity, charged particle. If it hits the DNA, it will damage it.

ANSWER: In order to determine whether a cosmic ray hit the DNA of an ancient bacterium, we need to know the cosmic ray flux and the size of the bacterium's DNA.

We are continually bombarded by cosmic rays. About one cosmic ray passes through a hand every second. Because a hand is about 4 inches (10 cm = 0.1 m) by 4 inches, that gives a rate of 10^2 cosmic rays per square-meter per second.[*] Cosmic rays damage molecules close to their path by "wiggling" their electrons.[†]

Now let's estimate the size of a bacterium. The smallest object we can easily see is about 0.1 mm (10^{-4} m) in size. Knowing that Leeuwenhoek saw "animacules" with his primitive microscope implies that cells are about 10^{-5} m. Bacteria are smaller than these, so their size is about 10^{-6} m, or 1 μm.

DNA provides the genetic information to direct the functioning of the cell. Let's estimate its size. Size, of course, is a very vague term. What we really want is the probability that, if a cosmic ray hits a cell, it hits the DNA within the cell. That means we want to estimate the cross-sectional area of the DNA. Some cells have nuclei containing their genetic material. This makes it easier to visualize the problem. The diameter of these nuclei are about 20% of the cell diameter (more than 10% and less than 50%, based on vague memories of high school biology classes). That implies that the cross-sectional area of the nucleus (and hence of the DNA) is $(20\%)^2 \approx 4 \times 10^{-2}$ of the cell area.

---

[*] Yes, I know a hand is larger than that. However this gives a nice, round number and is actually more accurate than using the correct hand size.

[†] "Wiggle" is a technical term. In laymen's terms, high-velocity charged particles passing through matter interact with the atomic electrons through electromagnetic forces, supplying enough energy to break molecular bonds.

Unfortunately, bacteria do not have nuclei. However, we can estimate the amount of DNA in the cell by bounding it. The proportion of the cell devoted to its genetic information, the DNA volume, must be less than 10^0 (100%) and more than 10^{-4} of the cell volume, giving an estimate of 10^{-2}. If the DNA takes up 10^{-2} of the cell's volume, then it must take up a bit more of the cell's cross-sectional area. Thus, this is consistent with the 4×10^{-2} of the area that we estimated for cells with nuclei.

Therefore we will use an effective DNA cross-sectional area of

$$A = (4 \times 10^{-2}) \times (10^{-6}\,\mathrm{m})^2 = 4 \times 10^{-14}\,\mathrm{m}^2.$$

At a cosmic ray rate of 10^2 per square-meter per second, that gives

$$n = R_{\text{cosmic}}\, A_{\text{nucleus}} = \left(10^2 \frac{\text{rays}}{\text{m}^2\text{s}}\right)(4 \times 10^{-14}\,\mathrm{m}^2)$$

$$= 4 \times 10^{-12}\,\text{rays/s}$$

passing through the DNA of the bacterium. This is not a lot of cosmic rays, but the bacterium has been around for a very long time.

In 30 million (3×10^7) years, the number of cosmic rays passing through the DNA is

$$N = nt = (4 \times 10^{-12}\,\text{rays/s})(3 \times 10^7\,\text{yr})$$

$$\times (\pi \times 10^7\,\text{s/yr}) = 4 \times 10^3\,\text{rays},$$

or four thousand cosmic rays. This means that there will be about four thousand broken chemical bonds in the bacterium's DNA.

Reviving any single bacterium is very, very unlikely. It will be much easier to reconstruct the undamaged DNA piece-by-piece from a sample of many bacteria.

Decaying protons

9.6

Some theories predict that protons are unstable and eventually decay into less massive particles, releasing radiation. Given that radiation from proton decay does not kill us, what is the minimum possible proton lifetime?

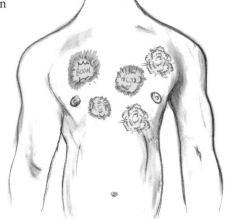

HINT: Protons make up 40% to 100% of all normal matter (by weight).

HINT: Each proton decay would deposit a reasonable fraction of its mass energy, $E_m = mc^2 = 1$ GeV, in our bodies.

HINT: If the proton lifetime is n seconds, then $1/n$ of the protons will decay each second.

ANSWER: In order to estimate the minimum possible proton lifetime, we should start with our maximum long-term radiation dose and then calculate the number of proton decays that would cause this, followed by the corresponding proton lifetime.

An acute radiation dose of 10 gray will kill us. An acute radiation dose of 1 gray will make us sick. We can presumably recover from that radiation sickness in under a year. This means that we should be able to absorb 1 gray each year. Although this will significantly increase our cancer risk, that is a longer-term problem. If we want to be more conservative, we can reduce all doses by a factor of ten, which will increase the corresponding proton lifetime by a factor of ten. A dose of 1 gray/year corresponds to energy deposited in our bodies of 1 J/kg-yr.

Let's only consider protons decaying in our bodies. Let's further assume that these protons deposit all of their mass energy in our bodies. This is reasonable because some of the radiation from protons decaying outside our bodies will deposit energy in our bodies, and some of the radiation from protons decaying inside our bodies will deposit energy outside our bodies.

Under these assumptions, the total mass energy of all the protons decaying in each kilogram of our body each year must be less than 1 joule. Because $E = mc^2$, we have

$$m = \frac{E}{c^2} = \frac{1\,\text{J}}{(3 \times 10^8\,\text{m/s})^2} = 10^{-17}\,\text{kg}.$$

This means that in every kilogram, at most $10^{-17}\,\text{kg}$ of protons can decay each year.

Thus, the lifetime of a proton must be greater than 10^{17} years.[*] That is 10 million times the age of the universe. Even based on just our fragile bodies, the proton must be a remarkably stable particle.

[*] The current lower limit on proton lifetime from sophisticated experiments is now about 10^{26} years.

If we travel by starship to the center of the galaxy, such that we arrive within a subjective human lifespan, how much radiation damage will we suffer from the interstellar "vacuum" (i.e., from the atoms in interstellar space)? Assume that Special Relativity holds, and ignore the problems of accelerating and decelerating the spacecraft.

¿ ¿

HINT: The interstellar vacuum contains about one hydrogen atom per cubic centimeter.

HINT: The solar system is about 3×10^4 light-years from the center of the galaxy.

HINT: Traveling close to the speed of light, our subjective time will pass more slowly by the relativistic factor γ. Each hydrogen atom we pass through will then appear to have total energy of $E = mc^2 \gamma$ as it hits us.

HINT: Each of those hydrogen atoms will then deposit energy in our bodies in the same manner as a cosmic ray, depositing 2 MeV for each centimeter of tissue it passes through.

ANSWER: In order to estimate our radiation damage from space travel, we need to estimate the number of atoms in our path and the damage done to us by each atom.

In order to estimate the number of atoms in our path, we need to estimate the distance traveled and the density of atoms. We might remember that the distance is about 10 kiloparsecs, or about 3×10^4 light-years. Alternatively, we might know that there are about 10^{11} stars in our galaxy and it is about 4 light-years (ly) to the nearest star. In that case, each star occupies a volume of about

$$V = (4\,\mathrm{ly})^3 = 200\,\mathrm{ly}^3,$$

so the total volume of the galaxy is $V = 10^{13}\,\mathrm{ly}^3$. The galaxy is relatively flat, so it could have a radius of 10^5 ly and a thickness of 10^3 ly. That would place the solar system about $d = 5 \times 10^4$ ly from the center.

Converting to meters, this means that the distance traveled is

$$d = 3 \times 10^4\,\mathrm{ly} = (3\times10^4)(3 \times 10^8\,\mathrm{m/s})(\pi \times 10^7\mathrm{s})$$
$$= 3 \times 10^{20}\,\mathrm{m}.$$

The typical density of interstellar space is one atom per cubic centimeter, primarily hydrogen. By comparison, room air has a density of $n = (6 \times 10^{23}$ atoms$)/(2 \times 10^4\,\mathrm{cm}^3) = 3 \times 10^{19}$ atoms/cm^3. If we consider a longitudinal slice of the spacecraft with surface area $A = 1$ cm^2, it will encounter 3×10^{22} atoms on its journey.

In order to travel this distance within a subjective human lifetime, our spacecraft must achieve a speed very close to the speed of light. Even traveling at the speed of light, if there is no relativistic time dilation, our trip will take 3×10^4 years, which is slightly longer than our expected lifetime. Fortunately, at speeds close to the speed of light, subjective time passes more slowly than it does for an observer watching from the center of the galaxy. We need time to pass about 10^3 times more slowly. In technical terms, this means we need a relativistic gamma factor of about $\gamma = 10^3$. This means that, from our point of view (i.e., in our reference frame) each interstellar atom will hit us traveling at almost the speed of light with the same gamma factor and will thus have a total energy of

$$E = \gamma mc^2 = 10^3 \times (1\,\text{GeV}) = 1\,\text{TeV}.$$

At these energies, when the hydrogen atom strikes the spacecraft, it will lose its electron very quickly. The bare proton will then pass easily through the spacecraft and our bodies, depositing energy in our bodies at a rate of 2 MeV per centimeter.[*]

Now we can consider the damage done by all these protons. Consider a volume of 1 cm^3 within our bodies. At the density of water it has a mass of 1 g. It will be hit by 3×10^{22} protons, with each proton depositing 2 MeV of energy. This means that the total energy deposited per gram will be

$$E_{\text{dep}} = (3 \times 10^{22}\,\text{protons/cm}^2)(2 \times 10^6\,\text{eV/cm})$$
$$\times (1\,\text{cm}^3/\text{g})$$
$$= (6 \times 10^{28}\,\text{eV/g})(1.6 \times 10^{-19}\,\text{J/eV})$$
$$= 10^{10}\,\text{J/g} = 10^{13}\,\text{J/kg}.$$

This seems like a lot, but let's compare it with a few things before jumping to conclusions.

As it only takes 2×10^6 J/kg to boil water, we would be vaporized more than 10^6 times over. In a thirty-year journey lasting 10^9 s, we would be turned to steam within the first 10^3 s, or 20 minutes.

[*] These protons can pass through about 1 km of metal before stopping. We can't stop them with mass shielding.

Unfortunately, we would not even survive that long. Radiation dose is measured in grays, where 1 Gy corresponds to an energy deposition of 1 J/kg. A lethal dose of radiation is about 10 Gy or about 10 J/kg. We will accumulate that dose in 10^{-12} of our journey, or in the first millisecond.

Who knew that vacuum could be so dangerous? I don't know about you, but I'm staying home.

Especial thanks to Dr. William Edelstein of the Johns Hopkins University School of Medicine for suggesting the idea and giving me permission to use it. See http://www.newscientist.com/article/dn18532-starship-pilots-speed-kills-especially-warp-speed.html for more details.

Appendix A
Dealing with Large Numbers

A.1 Large Numbers

We use scientific notation to write any number as a number between 1 and 9.99 (the coefficient) times 10 raised to some power (the exponent). Thus, we write 3 trillion as 3×10^{12} rather than as 30000000000000 (three followed by twelve zeros). Numbers with the decimal place on the right have positive exponents, and numbers with the decimal place on the left have negative exponents. For example,

$$0.0006 = 6 \times 10^{-4}$$
$$0.2 = 2 \times 10^{-1}$$
$$3 = 3 \times 10^{0}$$
$$2{,}000 = 2 \times 10^{3}$$
$$54{,}321{,}000 = 5.4321 \times 10^{7}$$

This has several advantages. First, as a professional scientist, I have lost the ability to count. It is much easier for me to write 10^{12} than to try to keep track of so many zeros.[*] We could ask a first-grader to count the zeros for us, but first-graders are not always available. Second, it reduces the likelihood of transcription errors. One of my students botched a problem because he misplaced a decimal point and wrote 11459.2 as 114592. If he had written 1.14592×10^{4}, it would have been much harder to make that mistake. Third, it draws more attention to the exponent, the most important part of any number. There is a much bigger difference between 3×10^{3} and 3×10^{4} than between 3×10^{3} and 4×10^{3}.

[*] As you might have noticed in the first paragraph.

Because we will be estimating, we will rarely (if ever), use more than one digit. Thus, we will round off 1.14592×10^4 to 1×10^4. This will make the arithmetic much easier. As discussed below, it will also avoid lying about (that is, drastically overestimating) how well we know the answer.

Now on to the arithmetic. When we add or subtract numbers, we adjust the smaller number so that it has the same exponent as the larger one. To do this, we shift the coefficient of the smaller number by one decimal place for each change in the exponent. For example, when we add 5×10^6 and 3×10^4, we increase the exponent of the smaller number by two and shift the coefficient by two decimal places so that $3 \times 10^4 = 0.03 \times 10^6$. (This is the same as writing 30 thousand (30,000) as 0.03 million.) Once the exponents are the same, we add or subtract the coefficients normally:

$$5 \times 10^6 + 3 \times 10^4 = 5 \times 10^6 + 0.03 \times 10^6$$
$$= 5.03 \times 10^6$$
$$= 5 \times 10^6$$

Given that we only work with one-digit coefficients, if the two exponents differ by two or more, then we ignore the smaller number. If the exponents differ by one, then the coefficient of the larger number will change by at most one (e.g., $3 \times 10^2 + 6 \times 10^1 = 3.6 \times 10^2 = 4 \times 10^2$). Addition and subtraction only really matter when the exponents of the two numbers are the same.

When we multiply two numbers, we simply multiply the coefficients and add the exponents. Thus,

$$(7 \times 10^6) \times (3 \times 10^4) = (7 \times 3) \times 10^{6+4}$$
$$= 21 \times 10^{10}$$
$$= 2 \times 10^{11}$$

Similarly, when we divide two numbers, we will just divide the coefficients and subtract the exponents:

$$\frac{7 \times 10^6}{3 \times 10^4} = \frac{7}{3} \times 10^{6-4}$$
$$= 2 \times 10^2$$

or

$$\frac{3 \times 10^4}{7 \times 10^6} = \frac{3}{7} \times 10^{4-6}$$
$$= 0.5 \times 10^{-2}$$
$$= 5 \times 10^{-3},$$

using the fact that $0.5 = 5 \times 10^{-1}$ to go from the second line to the third.

Note that we approximated 3/7 as 0.5. Actually, the correct answer is closer to 0.4, as $3/7 = 0.42857142857\ldots$ However, that is not worth worrying about. Remember that we are just trying to estimate closely enough to select one of the "Goldilocks" categories. If we worry too much about arithmetic precision, we'll never estimate anything.

Dare to be imprecise!

A.2 Precision, Lots of Digits, and Lying

In this book we do almost all of our arithmetic using only one digit (one significant figure). We do this to simplify the math and to avoid lying about our results. Both are important.

When some number in the news sets off our BS detectors, we should be able to check that number by estimation. We can only do that if we keep the numbers simple. At worst, we need a pencil and cocktail napkin to keep track of exponents and one-digit coefficients. At best, we would need a computer to calculate exponents and five-digit coefficients.

Yes, numbers with more digits are more precise. However, precision is not accuracy. Let's go to a rifle range. Accuracy means that the average location of all of our shots is in the bull's-eye; precision means all of our shots hit close together. If all our shots hit the bull's-eye, then our shooting is both precise and accurate. If all of our shots cluster in a dime-size hole far from the bull's-eye, then our shooting is precise but not accurate. If all of our shots are centered on the bull's-eye but evenly distributed over the entire target, then our shooting is accurate (on average) but not precise.

Because we are only trying to estimate within a factor of ten, our estimates fall, at best, into the third category of accurate but not precise. For example, we estimated that students spend between 2 minutes and 3 hours per day using their cell phones, giving a geometric mean of 20 minutes per day. As we would not be surprised if the correct average was 10 or 60 minutes, our estimate could easily be off by a factor of three. Therefore, it would be very silly of us to write

$$t = \sqrt{(2\,\text{min})(3\,\text{hrs})} = \sqrt{(2\,\text{min})(180\,\text{min})}$$

$$= 18.9737\,\text{min},$$

when none of the digits after the decimal point mean anything. Not only would we be wasting time, we would also be wrong.

Reporting lots of digits implies lots of certainty about the final answer. This is certainty we certainly do not have. For example, if a friend tells us that a nifty accessory costs $20, then we expect that the price is somewhere between $15 and $25, and we would bring enough cash. If our friend tells us that the accessory costs $19.17 including tax, then we could confidently go to the store with only a $20 bill, knowing that the cost was very close to the stated amount. More digits implies more certainty.

Similarly, a new tire might be advertised to have a 40,000-mile lifetime. That imprecision reflects the wide range of driving styles and driving conditions. It would be most peculiar to see a tire with an advertised lifetime of 38,769.3172 miles. That last digit implies that the tire's lifetime is known to within 6 inches.

Alternatively, suppose that we ask an astronomer how far the Whirlpool Galaxy is from the Earth, and she replies that it is 23 million and one (23,000,001) light-years away. When we ask how she knows the distance so precisely, she responds that many years ago it was 23 million light-years away, but that it is receding from us at high speed.[*] Her answer is very misleading; it implies that astronomers have measured that distance to extraordinary precision. In reality, the distance is only known to within 4 million light-years.

Thus, when we estimate a number, we will only calculate one digit of the coefficient. More precise calculations are both unnecessary and dishonest.

A.3 Numbers and Units

You may have noticed that I am using scientific notation and working in the metric system. The pizza box problem in the first chapter clearly demonstrated the advantage of the metric system. We converted the area of the pizza box from square meters to square kilometers. Had we worked in U.S. customary units, we would have had to convert from square inches to

---

[*] The Whirlpool Galaxy is actually receding from the Earth at a little more than 0.1% of the speed of light, so it would take 1,000 years for it to recede by an entire light-year. Please pardon the dramatic liberty.

Table A.1
Approximate Conversion Factors

Quantity	Metric unit	U.S. customary equivalent
Length	1 meter (m)	3 feet
Length	10^3 m (1 km)	0.6 miles
Length	0.01 m (1 cm)	0.4 inches
Area	1 m^2	10 $feet^2$ (10 square feet)
Area	1 km^2	0.4 $mile^2$
Volume	1 liter (L)	1 quart
Volume	1 cubic meter	250 gallons or 10^3 liter
Mass	1 kilogram (kg)	Mass of 2 pounds (lb)
Mass	10^3 kg	1 ton
Weight	1 newton (N)	0.2 lb
Speed	1 m/s	2 mph
Time	$\pi \times 10^7$ s	1 year

Note: These conversions from the metric system to U.S. customary units (and other useful conversions) are approximations. If you need more precision, look them up elsewhere.

square feet to square miles, using complicated conversion factors at each step.

The main difficulty with this, of course, is that most of us think in customary units. This means that we will need to convert from customary units to metric at the beginning of each problem and convert back at the end. To do this, we will need to know a few useful conversion factors, like the ones listed in table A.1.

Even with the headache of switching back and forth from customary units to metric, it is frequently much easier to do the calculations in the metric system.

Appendix B
Pegs to Hang Things On

These tables have been reprinted from *Guesstimation.*

Length in meters (m)	Object
10^{11}	Earth-Sun distance (1.5×10^{11} m)
10^7 (10^4 km)	Earth's diameter (8,000 miles, or 1.3×10^4 km)
10^6 (10^3 km)	Distance from New Orleans to Detroit (1,600 km)
10^5 (10^2 km)	Lake Michigan (length)
10^4 (10 km)	Mt. Everest (height)
10^3 (1 km; 0.6 mil)	George Washington Bridge
10^2	Football field (length)
10^1	Tennis court
10^0	Tall man's stride
10^{-1} (10 cm)	Person's hand (width)
10^{-2} (1 cm)	Sugar cube
10^{-3} (1 mm)	Coin (thickness)
10^{-4}	Human hair (thickness)
10^{-5}	Human cell (diameter)
10^{-6} (1 micron [1 μm])	Soap-bubble film (thicknes)
10^{-9} (1 nanometer [1 nm])	Small molecule
10^{-10}	Atom

Area in square meters (m^2)	Typical object
10^{14}	Land area of the Earth
10^{12}	Egypt; Texas
10^{11}	New York State; Iceland
10^9	Los Angeles; Virginia Beach
10^8	Manhattan
10^6 (1 km^2)	City of London
10^4	Football field
10^2	Volleyball court
10^0	Small office desk
10^{-4} (1 cm^2)	Sugar cube (one side only)
10^{-6} (1 mm^2)	Head of a pin
10^{-8}	Pixel on computer display

Density in kilograms per cubic meter (kg/m^3)	Item
10^{18}	Neutron star; atomic nucleus
10^9	White dwarf star
10^4	Lead; iron
10^3 (1 ton/m^3, 1 kg/L, 1 g/cm^3)	Water; human body
10^0	Earth's atmosphere at sea level

Mass in kilograms (kg)	Object
10^{30}	The Sun
10^{27}	Jupiter
10^{25}	Earth
10^{21}	Earth's oceans
10^{18}	Earth's atmosphere

Mass in kilograms (kg)	Object
10^{15}	World coal reserves (estimated)
10^{12}	World oil production in 2001
10^{11}	Total mass of human world population
10^{10}	Great Pyramid of Giza
10^{9}	Matter converted into energy by the Sun each second
10^{8}	Aircraft carrier
10^{7}	RMS *Titanic*
10^{6}	Launch mass of the space shuttle
10^{5}	Largest animal, the blue whale
10^{4}	Large elephant
10^{3} (1 ton)	Automobile (small)
10^{2}	Lion; large human
10^{1}	Microwave oven; large cat
10^{0}	1 liter or quart of water
10^{-1}	Human kidney; apple; rat
10^{-2}	Lethal dose of caffeine; adult mouse; large coin
10^{-3} (1 g)	Sugar cube
10^{-4}	Caffeine in a cup of coffee
10^{-6} (1 mg)	Mosquito
10^{-7}	Lethal dose of ricin
10^{-9} (1 μg)	Sand grain (medium)
10^{-12} (1 ng)	Human cell
10^{-27}	Neutron; proton; hydrogen atom
10^{-30}	Electron

Bibliography

[1] G. Bowley. U.S. markets plunge, then stage a rebound. *New York Times*, May 7, 2010.

[2] W. Poundstone. *How Would You Move Mount Fuji?* Little, Brown, New York, 2003.

[3] N. Carlson. 15 Google interview questions that will make you feel stupid. *Business Insider*, 2009. http://www.businessinsider.com/15-google-interview-questions-that-will-make-you-feel-stupid-2009-11.

[4] D. Adams. *The Hitchhiker's Guide to the Galaxy*. Pan Books, London, 1979.

[5] M. H. Abrams and G. Harpham. *A Glossary of Literary Terms*. Wadsworth Publishing, Boston, 2008.

[6] Crow calls for limit on loo paper. *BBC News*, April 2007. http://news.bbc.co.uk/2/hi/6583067.stm.

[7] Federal Reserve Bank. Money stock measures, 2010. http://www.federalreserve.gov.

[8] L. Weinstein and J. Adam. *Guesstimation: Solving the World's Problems on the Back of a Cocktail Napkin*. Princeton University Press, Princeton, NJ, 2008.

[9] Dr. Seuss. *The Cat in the Hat*. Random House, New York, 1957.

[10] NFPA 1710: Standard for the organization and deployment of fire suppression operations, emergency medical operations, and special operations to the public by career fire departments. Technical report, National Fire Protection Association, Quincy, MA, 2010.

[11] Ambulance fact sheet. Technical report, American Ambulance Association, 2004. http://www.the-aaa.org.

[12] Snopes.com. Piscine of the crime, 2000. http://www.snopes.com/science/poolpiss.asp.

[13] R. Iliev, S. Sachdeva, and D. Medin. Sinning saints and saintly sinners: The paradox of moral self-regulation. *Psychological Science*, 20:523, 2009.

[14] N. Mazar and C. Zhong. Do green products make us better people? *Psychological Science*, 21:494, 2010.

[15] M. S. Rosenwald. Why going green won't make you better or save you money. *Washington Post*, July 16, 2010. http://www.washingtonpost.com/wp-dyn/content/article/2010/07/16/AR2010071606839.html.

[16] Resin pricing—recycled plastics, 2010. http://plasticsnews.com/resin-pricing/recycled-plastics.html.

[17] *Statistical Abstract of the United States*. Technical report, U.S. Census Bureau, 2010. http://www.census.gov/compendia/statab.

[18] R. K. Iler. *The Chemistry of Silica*. Plenum Press, New York, 1979.

[19] C. Edwards and J. M. Fry. Evidence, life cycle assessment of supermarket carrier bags. Technical Report SC030148, Environment Agency, London, 2011.

[20] D. Williams, C. P. Gerba, and R. G. Sinclair. Assessment of the potential for cross-contamination of food products by reusable shopping bags. Technical report, University of Arizona and Loma Linda University, Tucson, AZ, 2010.

[21] Climate maps of the United States. Technical report, National Climatic Data Center, U.S. Department of Commerce, Washington, DC, 2005. http://www.ncdc.noaa.gov/oa/ncdc.html.

[22] Man-systems integration standards. Technical report NASA-STD-3000, NASA, Houston, TX, 1995. http://msis.jsc.nasa.gov.

[23] T. Robbins. The pedal-powered hotel. *Guardian*, April 14, 2010. http://www.guardian.co.uk/travel/2010/apr/14/hotel-with-electricity-generating-exercise-bikes.

[24] E. Fermi. Trinity test, July 16, 1945, eyewitness report. Technical report, RG 227, OSRD-S1 Committee, box 82 folder 6, U.S. National Archives, Washington, DC, 1945. http://www.nuclearfiles.org/menu/key-issues/nuclear-weapons/history/pre-cold-war/manhattan-project/trinity/eyewitness-enrico-fermi_1945-07-16.htm.

[25] Kenworth Truck Company. White paper on fuel economy. Technical report, 2008. http://www.kenworth.com/FuelEconomyWhitePaper.pdf.

[26] N. Kondamudi, S. K. Mohapatra, and M. Misra. Spent coffee grounds as a versatile source of green energy. *Journal of Agriculture and Food Chemistry*, 56:11757, 2008.

[27] N. Juster. *The Phantom Tollbooth*. Knopf, New York, 1961.

[28] T. P. Andert et al. Precise mass determination and the nature of Phobos. *Geophysical Research Letters*, 37:L09202, 2010.

[29] K. Koziol et al. High-performance carbon nanotube fiber. *Science*, 318:1892, 2007.

[30] V. F. Weisskopf. Search for simplicity: Mountains, waterwaves and leaky ceilings. *American Journal of Physics*, 54:110, 1986.

[31] A. Savage and J. Hyneman. *Mythbusters episode*, 2008.

[32] C. L. Cowan Jr., F. Reines, F. B. Harrison, H. W. Kruse, and A. D. McGuire. Detection of the free neutrino: A confirmation. *Science*, 124:103, 1956.

[33] Centers for Disease Control and Prevention. Emergency preparedness and response: Acute radiation syndrome. http://www.bt.cdc.gov/radiation/arsphysicianfactsheet.asp.

Index